The Asteroid Belt Almanac

SPRING 2014

THE Head & THE Hand

PRESS

EST. 2012

THE HEAD & THE HAND PRESS'S

The Asteroid Belt Almanac

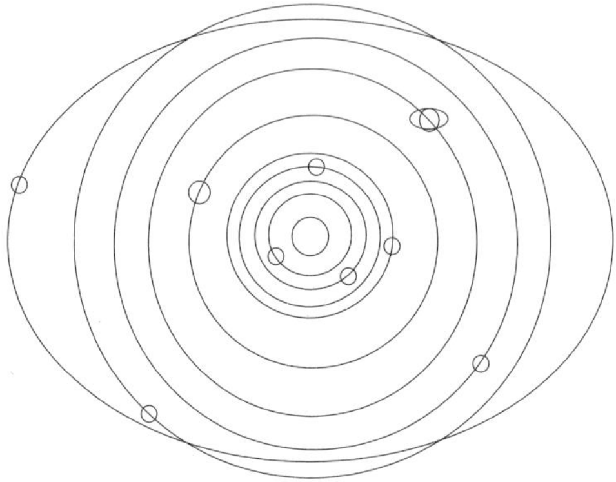

with the copyright year being **2014**

Of matters CURIOUS, USEFUL, and ENTERTAINING to the people of EARTH and BEYOND!

Cover art by Carpe Diem Construction Co. | Edited by Linda Gallant and Nic Esposito

Published in Philadelphia by The Head & The Hand Press | theheadandthehand.com

ISBN 978-0-9893125-4-7

Library of Congress Control Number: 2014934809

Table of Contents

Letter FROM THE *Editor*

Five months and three days ago, I received an email from an IP address originating in Philadelphia, Pennsylvania. I was in the midst of tracking the oscillation of light from a star in Galaxy M104, or as lay people like to call it, "The Sombrero Galaxy," which, by the way, I find somewhat offensive to the native people in the surrounding mountain town that I am barred from naming by the United States government (my de facto employer).

These oscillations, if proved to be consistent, make a very compelling case for the existence of orbiting planets surrounding the star. And to the overzealous observer, yes, they could make the case for life. Personally, I do not have the time for such abstractions, nor do I have the time for solicitations to distract me from my work. Before the advent of these nebulous and omnipotent communications systems, I would never have allowed a message to infiltrate the observatory. But after the third instance of my children and husband sending a government agent to break down the door to check if I was still alive after two days of relentlessly tracking the swirls of M51a, or the Whirlpool Galaxy, I realized that the occasional chirp of a message from the outside world was a necessary inconvenience.

The email issued from a publishing company calling itself The Head & The Hand Press and asked me to be the guest editor of its latest title, *The Asteroid Belt Almanac.* My first thought was, "How did these people get my contact information?" The second, after doing a quick Internet search for the company, was, "Why would an organization that conducts research on neither neurology nor orthopedics have such a name?" And my third and most perplexing thought: "Why, with an entire cosmos to explore, focus on the puny spectrum of asteroids between Mars and Jupiter?"

Despite my deep confusion and against my better judgment, I continued reading. Their proposal asked for my editorial assistance on what they called an "almanac," but which I understood to be journal of art, poetry, essays and fiction focusing on the sciences and technology.

I wondered, really, what could a group of artists contribute to scientific inquiry? Their letter reeked of pseudo-intellectualism and erudite-charlatanism. I immediately turned to their blog and book catalog in search of the keywords "Dianetics" and "L. Ron Hubbard."

When I was satisfied that they were indeed not Scientologists, I replied to their query, stating that I could be obliged to lend my services but under the condition that there would be no musings, no tangents, no pontificating. We would stick to hard data and irrefutable facts. They agreed, and thus the project began.

Upon reading the opening essay by the so-called "tech writer" Chris Wink, it didn't take long for me to consider holding The Head & The Hand liable for contractual negligence. As if the entire industry of pop-science reporting wasn't dubious enough, Wink's pining for scientific personalities to do more social good glibly disregards that a scientist's sole purpose is to seek the tangible truths of the universe rather than to solve the abstract ills of society. The collection went further off the rails with the subsequent story, "The Art of Cookery." Without any explanation, a young man from Cleveland wakes up in a cave outside of ancient Athens, suddenly fluent in Greek, and attempts to change the course of human history by retelling plots to movies from the 1990s. Hogwash. And the only profile of an actual scientist is in a story inspired by Justus von Liebig, a man whose claim to fame was introducing the world to an artificial meat substitute. I was in the process of contacting my lawyer before permanently canceling my webmail account when I came to "On the Outside Looking In," an essay by Sarah Grey.

Before I explain what eventually happened, I will offer the caveat that I have the maternal instincts of a praying mantis. The rearing of my children goes no further than tracking their grade point averages, and my relationship with my husband rarely extends beyond resource sharing and procreation. But this mother's reflection on advancements in medical technology working in tandem with the (objectively) astounding and intricate biological workings of childbirth gave me pause, however brief.

As Ms. Grey herself admits, she is not a scientist. But that did not prevent her from so eloquently articulating the lengths to which our obstetric medical technology can take us, as well as the limits humans sometimes choose to impose on those lengths. As someone who strives to look deeper and deeper into the ever-expanding universe, I was moved to find that we can learn so much by looking within ourselves. As I read back through those aforementioned stories, the ball of fury I felt at the thought of artists writing about science suddenly burnt up, like an asteroid entering our atmosphere. (I'm new at similes, forgive me.)

If technology promises to take us to the edge of human potential, why shouldn't Chris Wink thoughtfully reflect on where that edge may exist, a measurement I concede cannot be made empirically. If space and time can indeed be manipulated, then its manipulator would very likely experience the disappointment of young Stephanos in Mr. Huskey's timeless story. And finally, the cosmos, the boundless expanse to which I have devoted my life, may also house the expanded consciousness of artists and their art, as Mr. Hoots has posited in his essay, "The Music of Evolution."

I have done The Head & The Hand Press the service of reinforcing their collection with actual substance and factual information, but this was not an equal exchange. In return, they have given me the understanding that inquiry is furthered by a healthy dose of intuition, dare I even say dreaming. Tonight, I'm going to lock the door of the observatory, send word to my family that I'll be home by dawn, brew a strong cup of tea with extra honey, and sit here watching the Whirlpool Galaxy spin round and round.

Sincerely,

Celeste Teel

Dr. Celeste Teel

1 2 3 4 — Before Common Era (BCE)

c. 37,800 — Neanderthals create the world's oldest paintings in Spain's El Castillo cave.

2,000 — Rudimentary musical notation appears on cuneiform tablets.

c.1900 — A Babylonian tablet known as "Plimpton 322" lists sets of numbers that satisfy the equation now known as the Pythagorean Theorem.

367 — Aristotle begins studying at Plato's Academy of Athens.

10 11 12 13

1889 — Princess Margherita of Savoy, Queen consort of Italy, chooses her favorite among three traditional pizzas prepared for her by renowned chef Don Raffaele in Naples, Italy.

1922 — Pierre Teilhard de Chardin coins the term "nöosphere" in *Cosmogenesis*.

1946 — Yves Klein, Claude Pascal, and Armand Fernandez "divide the world" among themselves while sunning on a beach in Nice, France; Yves's designated domain is the sky.

1956 — Ultrasound technology is used for clinical reasons in Glasgow, Scotland.

22 23 24 25 26

2010 — Facebook develops the "Like" button as part of its Open Graph API project, designed to encourage third-party sites to share content and metadata.

2011 — Google debuts Conversation Mode for Google Translate.

2012 — Russian scientists drill into the largest subglacial lake in Antarctica, Lake Vostok, to study the ice's 400,000-year-old paleoclimatic record.

2013 — Royal Philips showcases its Philips ClearVue 350 portable ultrasound machine in rural Uganda as part of its "Cape Town to Cairo" tour promoting preventative screening.

2014 — The tenth anniversary of the Opportunity rover's landing on Mars.

Common Era (CE)

5 — **c. 150** — Claudius Ptolemy writes the Almagest, an astrological guide; its star catalog lists forty-eight constellations

6 — **1610** — Galileo observes the four moons of Jupiter and dubs them the "Medician planets" in honor of the Medici family.

7 — **1824** — Ludwig von Beethoven completes his Symphony No. 9 in D Minor.

8 — **1833** — An estimated 1,000 shooting stars per minute blaze over the North American continent during the Leonids meteor shower

9 — **1865** — Justus von Liebig founds the Liebig Extract of Meat Company.

14 — **1958** — Using satellites equipped with Geiger-Muller tubes, James Alfred Van Allen discovers radiation belts held in place by the Earth's magnetic field.

15 — **1963** — Madeleine L'Engel delivers her Newbery Award acceptance speech for *A Wrinkle in Time*; it is entitled "The Expanding Universe."

16 — **1978** — Brian Eno's Music for Airports, the first album in his "Ambient" series, is released.

17 — **1984** — Kurzweil Music Systems releases the first synthesizer, the K250.

18 — **1987** — The first presentation of GIF image formatting depicts an airplane.

19 — **1990** — The Hubble Space Telescope is launched into Earth's orbit, moving at a speed of five miles per second.

20 — **1993** — The first issue of *Wired* magazine hits the stands.

21 — **1999** — Jim Buckmaster, computer programmer and the creator of "Missed Connections," is hired by Craig Newmark after posting his resume on Craigslist.

Share Something Greater

Christopher Wink

Sometime after one in three humans on the planet started using the Internet, the speed with which we changed our lives began to slow. Widespread solar paneling and 100 mpg automobiles are still nobody's normalcy. We can't seem to look up from that text message we're writing on our increasingly crowded smartphone to look toward bigger horizons.

What I'm asking you is to set aside your crowdsourced social network spin-off. Put down your daily deals discovery mobile app. A generation bred in a transformative era for communications needs to chase new challenges. So why are we still so focused on the ones we've already solved?

This is nothing new, of course. Technology follows a predictable path.

Someone creates something inventive enough to be called magic. It's too expensive to buy, so for a time we just marvel at it behind store windows. But then someone else figures out how to sell it and people get wealthy doing just that. Then over time, a steady parade of efforts come along modifying or leveraging that invention to make successively less impact and less money. Finally someone starts the process all over again.

A generation ago, that process began when we were thrust down into the age of networks. In the same way that the completion of the transcontinental railroad sped the American expansion westward 150 years ago, in recent decades we have laid the groundwork for an increasingly global communications system, lubricated by automation software and digital discovery. It's an effort that began with the noble goal of ensuring that all of human knowledge can be accessed, annotated, and distributed, and now we're building mobile apps that put quirky filters on photos we take at parties.

The infrastructure needed for that noble goal has already largely been constructed, but we continue to tweak and refine it without realizing the true social change that the age of networks was meant to deliver. Instead, the very best of my generation are working to perfect the roundness of the "Like" button.

We've come to a moment when we can either linger or find a greater good.

We put cell phones in the hands of people experiencing homelessness, tablets with children in dismal schools. But we've spent so much time building that information highway that it seems we have meandered from what the purpose was meant to be. Today, we have geolocation tools that recommend restaurants rather than predict crime to help prevent it. We have built just-in-time logistics management software that can maximize efficiencies for picking up dry cleaning but not yet for distributing potable water.

There is nothing wrong with tinkering with what has already been built to benefit those of us at the luckiest end of the spectrum. Trickle-down innovation is inevitable. Privatizing space travel to usher in a new age of competitive scientific discovery will also eventually ensure that we will someday have all-inclusive resorts on the moon. Likewise, someone will commercialize affordable personal genome mapping, and then someone will use that advancement for a sweet online dating app.

But we need loud leaders whom we cherish and celebrate for doing more.

If the age of networks has given way to this, the social age, then the battle of our time will be whether that age will be remembered for lingering triviality or foundational change.

It very well may always be cyclical. We make our lives more comfortable, profit on that change, get distracted in less relevant work, fall into hardship, and then build our way out to start it all over again. In the end, like any great, grand platform, we get the technology and the society we deserve.

ON THE **MISUSE** OF THE **TIME-SPACE CONTINUUM**

The Art of Cookery

Wintfred Huskey

One morning Steven Summers woke up in a dark cave somewhere in ancient Arcadia. Steven didn't know it right away, but outside of the cave a "Golden Age" of Athenian hegemony was busily going on throughout the Mediterranean, which meant it was a pretty good time to be in Arcadia, too. The whole region was on the up and up. So if you were to find yourself transported to an ancient cave, you could do *much* worse. But the funny thing was that Steven went to bed the night before in his apartment outside of Cleveland, Ohio, in the year 2013. Don't ask him how all of this happened, but it did.

He was understandably frightened and confused when he woke up inside of his dark hole. Cautiously making his way out from the cavern, he winced in the blindingly bright light. Steven patted at and slowly noticed his new old-fashioned garb: a tunic and sandals. The mesh Adidas shorts and rec league softball shirt he went to bed in were no more and there was nothing else of his on his person or where he awoke in the cave. He stood outside and savored the scene for a moment anyways, like a proud homeowner might while fetching the paper from his manicured lawn. It was a nice day, maybe springtime. The greens and blues were cheerful, naturally fluorescent. His outfit was breezy and comfortable. It had an undeniably relaxed feel despite the inherent strangeness of the situation.

As has been commonly experienced by humans in moments of distress throughout history, the happy chirping of birds assuaged and pleased Steven.

Walking along a nearby path, Steven spoke to the first person he encountered and was politely told where he was. His native English had been swapped for Greek, which was surprising, useful, and profoundly reassuring.

Steven got over his apprehensions and fears quickly because when compared with the bucolic landscape and adventure that Arcadia welcomed him with, life outside of Cleveland was pretty ugly and dull. In fact, he wished and daydreamed quite often about this sort of thing happening. Up until last night, Steven lived in an apartment, alone, and didn't miss anyone in particular from the future. He'd taught sixth grade for the past three years—at the same school he'd gone to as a boy, no less. His mom even used to be a secretary there.

Steven, by the way, didn't feel bad about his parents not being in Arcadia with him because he knew that they would come along years from now, meet one another, and eventually be retired in the future. They would one day live in white-sanded Florida comfort, with ample monthly benefits. Steven didn't know if the same could be said for himself.

Remembering what he did from *The Odyssey*, Steven planned on wandering around and receiving bread and wine and the odd hunk of roasted meat because Hellenistic cultures had this thing for supplicants. Not only did they not mind bathing, anointing, and feeding total strangers, but they seemed to *like it* (especially if the wanderer could tell the local court a thrilling tale over a dinner feast). And there were all sorts of stories he could entertain people with. In fact, all Steven had to do was think of stories he knew that weren't written yet (*i.e., most stories he knew*) and just change the names and make things take place in the Mediterranean.

Suddenly he was quite excited to have woken up in that cave.

But at that present moment it was breakfast time and Steven was pretty hungry. Even the greats needed a good meal to start the day. His head hurt a little. Of all the drinks not yet invented, he'd unquestionably miss coffee the most. *Or*, Steven thought to himself, *better yet: a drink for me to invent*! He'd have to get a line on some beans.

The first house he happened upon was a modest rock hovel—a dwelling not without dignity or charm, however. He walked up to the goatskin hide covering the entrance and gave it a 'shave-and-a-haircut...*bomp-bomp*' knock. A young, dark-haired boy answered the pelt.

"Good morning, young man. Is your mommy or daddy home?"

The boy nodded slowly.

"Can you tell them that there's a supplicant here to see them?"

The boy nodded again and disappeared inside, letting the skin flap back down over the entrance. Then came some murmuring. Steven patiently waited and whistled "Yesterday," which he made a mental note to write down and perform someday, as he knew both the words and chords.

A much larger, more bearded version of the little boy pulled back the goatskin and stared uneasily at Steven. He wore a mean expression and a rough leather jerkin.

"Can I help you?" the man growled.

"Yes, hello, my name is, um, *Stephanos*, and I'm a supplicant. I thought I'd stop by for something to eat this morning and maybe tell you and your family a story."

The man had a decidedly modern reaction to Stephanos's offer. "We haven't any bread," the man said, "now move along, filthy beggar."

"But I thought Zeus said–"

"Listen: Zeus does not make the rules in *this* house. Theophylaktos makes the rules this house. I am Theophylaktos and I command you to leave, lest I begin flogging you. Do you not see that I am busy? That I must go check on my sheep? Now on with you, you licker of pots!"

He slammed the goat hide in Stephanos's appalled face.

...

Stephanos was earnestly hurt by Theophylaktos's name-calling. He'd always thought of the Greeks of old as a remarkably hospitable people. He could imagine this sort of thing going on in *Rome*, maybe, where every manner of meanness went on, but never in Greece. He trudged on, more than a bit disheartened, and tried to ignore his hunger by thinking of all the mundane common knowledge he could impress people with. But the sun was hot, high up in the bright blue sky, and he surely needed a drink. It would take a cool river to slake his thirst. Eventually the road Stephanos walked down passed by a rather stately home, smaller than a palace, though. An estate, you might call it.

He knocked and waited before the grand wooden door. He hunched over with his tongue cartoonishly sticking out of his mouth, panting. He was actually that thirsty.

The woman who answered said she'd have one of her handmaidens fetch some water and light snacks for the weary traveler. Stephanos was lead to a rolling green courtyard and instructed to wait there. Near him a young boy and his tutor were going over a math lesson.

"Now, *boy*, let us see if you are not as dull as you've been letting on," the tutor said. One glance at the tutor told you that he prioritized matters of the intellect over vulgar, illusory things like physical appearance and personal hygiene. "Here is a rather simple one: if two magnitudes be commensurable, and the one of them be incommensurable with any magnitude, the remaining one will also be incommensurable with the same." The man raised his eyebrows like a pair of arching caterpillars. "Go on, then. Prove it, you waggish rascal."

The student could not have been more than nine or ten. He was as yellow-haired and friendly as a television family's beloved golden retriever. "Alright, then, sir. Let A and B be commensurable magnitudes, and let A be incommensurable with magnitude C."

"Very good, go on."

"The remaining one, that's B, will be also incommensurable with C.

"And why is that?" the tutor prompted the boy, stroking his billy-goat chin.

"Well, sir, if B is commensurable with C, while A is commensurable with B, A has to be commensurable with C."

"And how is it that you *know* that, my young friend?"

The boy seemed stumped. "Um..."

"Come now, really! This is quite simple!" the tutor shouted. "Which proposition proves what you just let pass through your impudent, loutish lips, you donkey-eyed dullard?!"

The boy winced. "Proposition...*four*?" he guessed.

"Four!? Four!?" The tutor's voice rose into a nasal screech.

He swatted the boy with a small baton he clutched, hidden in his hand the entire time. The thwack it made against his head left an immediate strawberry welt. "You twit! You infantile buffoon! You haven't paid any attention at all! You waste my precious time! You tell me that you arrived at this conclusion because *given three commensurable magnitudes, one can find their greatest common measure!?*" Dry spit was firing out of his mouth like corn kernels popping. "No, we just went over this, you *ninny*! *Proposition 12: Magnitudes commensurable with the same magnitude are commensurable with another also.*"

The boy's quiet blubbering and slow-trickling tears simmered the tutor down a bit. "Come now, master, this is common sense." The tutor took on a high-toned, conciliatory posture. He straightened his toga, which had begun to slip off his shoulder. "You must pay close attention. It's for your own good. Let us finish the proposition and move on to the next one before our lessons conclude for the day."

The boy sniffled his tiny nose and finished up the problem correctly. Stephanos, meanwhile, sat stunned, not noticing the jug of water and figs brought to him. He was supposed to start teaching his students how to divide fractions today, he remembered.

...

The day did not cool down once Stephanos returned to the road. His sandals provided little in the way of arch support and in his mind he concocted a design for a rudimentary desert boot. Stephanos would soon need some papyrus and a writing instrument, lest he forget his many great improvements.

By the early afternoon his feet hurt so badly that he could not walk any farther. He felt flat worn out. He found a nice tree near a river and sat in its peaceful, leafy shade for a rest. Licking his lips, he was ready to bring pizza to the ancients, thousands of years ahead of schedule. Perhaps he could even settle on hanging his reputation immemorial on just that: being the creator of pizza. Would he not be as fondly remembered and as beloved for having done that as, say, introducing the place-holder '0' to Western civilization?

While he was considering what barley flour crust would taste like and whether or not yeast was available, the sound of drums and woodwinds came drifting in on a gentle Arcadian wind. It sounded a lot like an early Genesis album. Soon the music was accompanied by the smell of burning wood and a distant hubbub. Stephanos got up and reconnoitered from behind a shrub at the top of a hill.

A small party was getting started in the dale below. Three or four people scuttled about while a circle of half-a-dozen men stood and talked. The fire had just got going and a butchered animal waiting to be roasted lay on a pile of leafy branches and twigs. There were three great earthen vessels in iron stands and a long table covered with various bowls and dishes. Some far off day the dinnerware he spied might be displayed in museums around the world.

Stephanos cagily peered from afar and waited for the perfect moment when the crowd got big enough for him to slip in unnoticed. Then he could get some free food and booze. This Greek get-together wasn't so different from a frat party. *Maybe that's why they kept the letters,* he thought. When the meat was skewered on a spit, he could wait no longer, though. An irresistible wisp of smoke beckoned him. He casually walked towards the festivities.

He grabbed a red clay mug and walked over to the large vessels.

"How you doin'?" he said to a wiry old man holding himself upright with a staff. The man grunted in crotchety assent. Stephanos removed the cork from a spigot and olive oil streamed out. "Whoops," he said, corking it back up and filling his mug up at the next, correct, wine-laden crock before circulating.

"What kind of meat we got there?" he asked the pit master.

"Oxen."

"Mmm-mmm. Roasted oxen. That's Apollo's favorite. Mine, too." He took a sip of the wine and nearly did a spit-take (which wasn't considered in the least bit humorous yet and simply seen as an indecent medium between belching and vomiting). The wine tasted awful, like vinegar cut with molasses. But he swallowed it down. Stephanos perused and ambled. He made a full circuit of the scene after maybe four minutes. Finally he settled near a group of eight bearded men in the midst of a very lively discussion. Stephanos listened as one man, bald with a woolly scruff like a sheepdog's coat covering his face, was in the middle of a thought.

"Of course, we know there to be seven bodies having seven orbits in our night sky, do we not?" The crowd agreed with a collective hushed breath. "Let us consider the four closest. The moon in the first lane, and the sun in the second, hovering above it; then there is the Dawnbearer, so sacred to Venus, and the star of Hermes, both of which also glide about at the same pace as the sun." He paused and allowed the crowd time to take in his words. "Now. We also know that the sun, the star of Hermes, and the Dawnbearer all over*take* and are over*taken* by one another. If we then notice that the other bod—"

"Excuse me," Stephanos interrupted, having just gulped down another rancid mouthful of wine, "but have you ever considered that the *bodies*, us included, actually revolve around the sun, which is fixed in the middle of things?"

He expected rowdy laughter which he would smugly wait out only to then gall them with an eloquent explication of the heliocentric universe. But instead everybody sat uncomfortably for a moment. Had phonographs been around, a needle would have screeched to a halt. Stephanos's words must have been uncouth, decidedly *au gauche*.

"Sorry, I don't mean to intrude, it's just that..." He felt suddenly nervous, totally unable to describe a concept he taught his students in science every February. He laughed to himself in a weak attempt at personal composure. "So, like, basically, what I'm saying is: what if *the sun* was in the middle, right? And all the other planets circled around it, you know? Don't you think that would be a lot *simpler* than making all this stuff spin around *us*, with all that retrograde motion and, you know, *stuff*... Somebody get me a stick so I can draw it out in the dirt."

Nobody moved to get him a stick or seemed particularly impressed, least of all the bearded one who was talking before he horned in.

"Young man," said a brusque and squat listener wearing a buttery-hued toga. He stood up directly next to Stephanos. "This idea is not a new one. It is merely a crappy one."

"Well—"

"No, stop there, that is quite enough. Your feeble and

frankly insane sophistry is a waste of our collective time. Suppose we *are* moving about through the heavens—a ludicrous thought, but let us consider it nonetheless. How, then, can we also be sitting still?" The crowd murmured. The man offered a self-congratulatory smile to his friends as he paced in a slow half-circle until he was across from Stephanos. His voice was brassy and he had an appropriate hand gesture for every point he made. "Were we actually jostling about through space, how, then, would our noble temples stand firm and proud? ... And how would the tripod of sweet, murky wine stay upright and not topple over, causing no uncertain revelry amongst the ants and snakes? ... My friends: how would the wine-dark seas remain so calm and placid? Does not the water-full pot *splish* and *splash* when we try to carry it about?"

The listeners clapped deferentially, as if he'd just sunk an easy put on a golf green. The silver-tongued cynic impressed the group, making Stephanos feel impotent and small. "Well," Stephanos recovered, "this might sound kinda *weird*, kinda *kooky*, but there's this force called *gravity* that keeps things together, keeps small stuff from flying away from bigger stuff...bu-ut, the thing is, you can't see it. Like I said, it's more of a *force*."

This explanation was woefully lackluster and indeed kooky. The ensuing silence was, in a word, hostile. Someone mercifully blew a horn announcing that dinner was ready, however, allowing Stephanos an escape. He seated himself off to the side, far away from the priggish men he'd tried to explain the universe to.

He was quite embarrassed and hungry, so he tuned out the host's preamble. He didn't particularly care what they were celebrating. While chewing a hunk of oxen, Stephanos thought deeply about what he knew well enough to explain. So much of what he took to be true involved things he couldn't see or hadn't been present for, or things that he'd only seen pictures of, or that he'd never actually thought too carefully about. He knew all about the benefits of handwashing, for instance, but could he convince these curious men to keep their dirty fingers out of a wound? It was doubtful—they'd pepper him with a bunch of smart questions and incredulousness about invisible microbes and germs.

Suddenly he realized he was being addressed. "Stranger, we welcome you at our table, as we are learned men, men who embrace ideas and thought. We are most magnanimous men." He paused. "That is to say, most of us are magnanimous men—I'm looking at you, Lysimachos!" Everybody laughed, gently teasing Lysimachos. "But, seriously, please, do tell us something about yourself." The man speaking had a rich, velvety baritone. His benevolent presence calmed Stephanos.

"Well, thanks everybody for having me. My name's Stephanos and I hail from...Thebes. And I'm a...traveler. I heard your merriment and figured I'd pop in and, you know, check things out." He looked around worriedly. The element he'd spoken to earlier still glared disapprovingly, but everyone else seemed slightly buzzed and softly grinning. "And since you all have been so gracious to me, especially with my being a supplicant and all—"

"As Zeus himself commands!" someone yelled out.

"Yes, *thank you,*" Stephanos continued, "as an appreciative guest, I thought I might entertain you with one of our favorite Theban tales."

The crowd looked to one another and nodded excitedly. In muted tones they voiced intrigue over the stranger's offer. Stories about the freaky things happening in Thebes were quite popular then.

Stephanos began the invocation.

"*Sing in me muse, and through me tell the tale of Kevin—*"

"Kevin?" someone audibly whispered.

"Shh. It must be one of those queer Theban names," explained the kingly host under his breath. "Go on, Stephanos, do continue!"

He smiled smugly at the crowd. "Thank you. *Sing in me muse and through me tell the tale of Kevin, the young lad of many ways, who was left home...alone! Yea, while his family journeyed far away without him; after their indifference and boorishness had made Kevin to* wish *them away. His wishes were granted by Zeus of the deep-furrowed brow, and many were the pleasures in which he partook, the treats he ate; but many, too, were the lonesome nights when he missed his mother and father and even his oafish brother Buzz and vile uncle Frank. Even so, he remained firm in the face of danger, ever courageous and ingenious, for the bumbling boobish burglars were to be beaten by a boy, subdued by a golden-haired youth of seven. From here, goddess, wise and venerable daughter of Zeus, speak and begin our tale.*"

Stephanos retold *Home Alone* much to the delight of the crowd—and irrefutable proof that the film was a timeless classic. He played to local partiality and made the thieves Persian and changed the old man with the snow shovel into Athene, all of which the revelers ate up. By the time Kevin's family returned, everybody was quite drunk and festive. Stephanos was himself intoxicated by not only the now-delicious wine, but also the hearty applause and rowdy laughter.

"Your unique brand of comedy was most satisfying, Stephanos!" intoned the head of the party, wiping away a cheery tear. "Perhaps you could favor us with another story?"

"Yes, something more *dramatic*!" another voice shouted. "You know, a *Theban* drama!"

Everyone responded with a ribald and suggestive '*ewwwwww*' noise before pounding on the heavy wood banquet table. Stephanos stood patiently, smiling rather coyly, as though he had a surprise in store for them.

"Very well, gentlemen. I must warn you, however, that this tale is a cautionary one—some might even say a *horrifying* one."

The crowd was completely silent. Nobody had thought to tell horror stories yet. They only had dramas and comedies at this point in time.

But his rehashing of *Jurassic Park* was doomed from the get-go. The audience found the idea of enormous terrible lizards roaming the earth before mankind to be dubious and borderline blasphemous. Getting past that, they could not understand why an eccentric tycoon would want to bring back such creatures in the first place. Stephanos never thought of that, either, he had to admit. And, besides, how *did* he manage to bring the lizards back to life? There was science behind *Jurassic Park* that he remembered fairly well and tried mightily to explain: the cloning, the genetics, the bug in the amber. After that fell short, however, he concluded that magic and sorcery were involved.

"All right, Stephanos, thank you," said the host at this juncture. "Your...*inventiveness* will surely be remembered by all. Now—bring out the desserts!"

The guests cheered, over-stuffing themselves with sweets like only drunks and children can, but Stephanos sat sullenly and brooded. His barley cake was dry and flavorless and the honey slathered all over it made things harder to swallow.

...

The next few weeks were increasingly frustrating for Stephanos and soon he began wishing he'd wake up in Steven's reliable, normal Cleveland. Making it as a poet was less easy than he anticipated. There were few stories for his ancient audiences that he actually knew all the way through. He once tried to recount *Romeo and Juliet* (he taught a school-district-approved version of the play each spring to his class), but the plot was incredibly complex and difficult to remember fully. He'd always mix up the delicate order of the story or have to back up and explain something that he forgot to mention. And whenever he did tell the tale perfectly, the listeners found the multitude of characters, repeated coincidences, and main themes wholly outrageous.

Once, upon seeing a beautiful woman in a marketplace—a veritable Grecian goddess, the sort of perfectly plump and milky type that painters would obsess over 1900 years from then—Stephanos recited some Yeats. He stared deeply into her lovely eyes, green and serene, the color of a bottle of fancy mineral water, and asked her to tread lightly, for he had spread his dreams beneath her feet. She took him to be simple and potentially dangerous and politely thanked him before she urgently implored him to "look there!" and bolted off as he dumbly stared off in the direction she had pointed.

Stephanos tried his hand at music for a while, too. He wasn't particularly gifted with the lyre, though, and his attempts at becoming a luthier resulted in something more like a milk-carton ukulele than a guitar. "Excellent," he said, applying the finishing touches to some spiffy tuning pegs he'd jerry-rigged, "now to put on some guitar strings." As another painful realization washed over him, Stephanos began to cry softly to himself. He'd never felt so useless in all of his life.

For the next twenty days Stephanos wandered around Attica, taking in the sights and occasionally trying to offer some helpful twenty-first century advice to whoever would listen. He couldn't get people to cover their mouth when they coughed or sneezed, though. Nobody would believe his predictions and occasionally dire warnings (which were, in defense of the ancients, only half-correct, as Stephanos had gotten a C+ in the World Civ course he sparsely-attended in college) and few were the people that would pay any attention to his far-out theories on physics and chemistry. The ones that did listen were usually just argumentative types.

For example, Stephanos would gather a small group in a public space and declare something like "People! Harken! Everything in the universe is made up of tiny particles I call atoms."

"What was that you said? These 'atoms,' they make up everything?" a doubtful man would ask with an interested look on his face, the leery member of an infomercial audience.

"Yes, they are too small to see, but they bunch up together and sort of, well, *cluster*...like your average bunch of grapes, in fact. Yes, like grapes. Those clusters then become what I call *elements*. Most of you seem to think there are only four of these, but there are actually dozens. They team up and make each and every thing surrounding us."

"Fascinating! So what you describe is how, say, that tree is made?"

"Yes, correct," Stephanos would reply. "The tree is mostly made of a common element I call 'carbon.'"

And the people would snicker at the air-quote gesture Stephanos did when saying "carbon." "And even swift-footed Achilles? He was made of these atoms and elements of which you speak?"

"Yes, precisely," Stephanos would say excitedly. "Much of the same stuff, in fact!"

"And the batshit crazy words you're saying, too? They're made up of atoms?" And he'd pause for laughter. "Also made of 'carbon,' no doubt?" He'd barely manage to get the last words out while choking on his own chuckles after a deriding doff of air quotes.

"No, those are sound vibrations, and that's rude and beside the point. Atoms are—"

"Explain fire," another person would say. "Yeah, do fire!" another would jump in, causing Stephanos to wearily shake his head.

Whenever this confusion happened to his class in Cleveland, he could use the Periodic table and marshmallows and toothpicks and then even the dumber kids would grasp the concept. But he didn't know that whole goddamn table by heart and he certainly didn't have any marshmallows.

And, oh, how he'd kill for some marshmallows. The food in the past was terrible and it was getting to him. Stephanos now completely understood why wars were fought and men died attempting to get salt. Things he ate tasted, at best, like nothing. His immense foreknowledge of future recipes was not enough to overcome the ingredients that were at hand and the available kitchen technologies. So everyday he'd grumpily shuffle from village to village and retell a story (which, usually, he didn't care for, a movie like *Titanic* or *Toy Story*) and then he'd get to eat enough barley-meal and boiled roots to sustain himself and do it over again.

...

Like most people painfully aware of their burdensome genius and superiority, Stephanos felt ignored, rejected, and generally embittered by the world surrounding him. Did they even deserve him? It wasn't like some complete schmuck for some unknown reason had been sent back in time. He had been in school for twenty of his twenty-nine years (six of those at The Ohio State University) and he never got lower than a C+. As an accredited Master of Education he was required by state law to be knowledgeable about a variety of topics, and he had test results proving that he was more than adequate in math/science, and composition/reading comprehension. But he could neither prove nor demonstrate most of what he knew to be true. The Greeks were, by and large, too hard-headed and difficult to impress.

So although Stephanos admittedly had no true skill or craft that he could practice or make a living at, he could not help but feel strongly entitled to a life of greatness and recognition. Naturally, he decided to strike out for the city, Athens, where a cultured and far-seeing man might yet have a shot.

The first thing that won him over about Athens was that you could get a decent meal there. The wine was still pretty awful, but they at least practiced the art of cookery. Coarse salt, briny olives, tangy cheeses of undetermined milks; oil, vinegar, fragrant herbs; figs, beans, rich roasted acorns...and they ate in comfort, too, while lying on couches or at fine tables, sitting on luxuriously upholstered stools. The country had its quaint straw beds covered with myrtle and it had its rustic loaves, too, but the city did it all better.

Stephanos fell in with a group of philosophers who were loosely loyal to several different pre-Socratic thinkers. Among their many beliefs, they too posited that the universe was composed of many tiny pieces. Most days and evenings they hung out and tried to quilt together various far-flung and occasionally oppositional concepts into a single, coherent doctrine. Stephanos liked their pragmatism and knew that, of all the academies and factions he'd encountered in Athens, these guys were closest to the mark. Best to stick with them. And best yet, one member of the clique, Diododos, was fabulously wealthy and took very good care of his colleagues.

City life distracted Stephanos with its wiles and he became less and less focused on becoming a bold-faced name in history books. A few weeks in Athens helped him to decide that it had been foolish in the first place to try and revolutionize an entire society like an enlightened tornado. It was necessary to bide his time while getting his various thoughts organized and on papyrus. It's what separates the luminary from the crank. Only occasionally, when drunkenly discussing something with his fellow philosophers, would Stephanos let slip some future truth. And this was always blurted out as nothing more than a vague, leading statement before he'd hang back, happy to nudge the discussion along towards what he knew was the correct answer. He fondly remembered doing this all the time with his students, how it developed their budding self-esteem, thinking that they "got" the message behind, say, *The Giver* (even though Mr. Summers had given them many hints leading them to a designated conclusion).

As the months went on, Stephanos became something of a leader amongst the members of the academic troupe he palled around with. The group gradually earned a reputation within Athens. After a few months they even had a growing number of followers. Rival gangs of philosophers would come and challenge them in debate, like opposing crews of break dancers. Stephanos and Co. usually came out sounding smart and deftly proved their points, winning over more disciples in the process. Secretly, Stephanos finally began working on his treatise during his spare time and also began serving unleavened bread with melted cheese at the inner-circle's meetings to great acclaim.

One day the team got together at Diododos's house for a dinner party-*cum*-conference. They were supposed to continue fleshing out their stance on whether or not the void (empty space) existed or was, by its nature, non-existent. But the talk kept turning to politics and war. Stephanos, for one, was tired of the diplomatic gossip and fearful chatter.

"Gentlemen," he said, his voice rising above the clamor of his peers, "we ought to be above these superficial diversions. Politicians will always be stupid and corrupt and desirous of war, which will never be vanquished." Nobody said anything. "I mean, sorry to be a downer, but that's the truth. I know it *for a fact*."

"Shouldn't we seek governmental offices then?" offered Chandros, a young, tubby acolyte. "Wouldn't the surest way to improve our society spring forth from the legislative process?"

Everyone agreed. Stephanos forgot how annoyingly devoted the Athenians were to their legal system, their citizenship, and their laws. "No, that would only be a waste of your time. Believe me: democracy is *a bit* overrated. You should concentrate on the intellectual work we're doing here."

To wit, Chandros replied that he'd be busy concentrating on not becoming a Spartan slave. The crowd overwhelmingly approved. Stephanos grabbed his things in a huff and took a mug of wine to go.

"Fine, just wait and see if I'm not right!" he exclaimed ruefully. "Don't come crawling back to me for answers when I'm famous and you're all reading my book!"

So for the next month, Stephanos worked feverishly on his treatise, suddenly with a real purpose (to rebuff his old friends). His final pile of sheets was on the thin side but managed to cover basic biology and health science, some general physics (with far-reaching implications), and, among other scattered thoughts, included twenty-five basic, easy-to-prepare recipes. The book would beef up once some helpful illustrations were added (Stephanos couldn't draw for sour apples). He went outside infrequently during this time, and then only to get something to eat or drink. He was immersed in thought and, when not writing, tended to always look straight down at his feet, mumbling to himself on the way to the baker or the stream. Mothers wearily pulled their children close whenever he passed by.

By the time he put the finishing touches on his rough draft, his sense of accomplishment was tempered by a much more literal sensation of fever. Perhaps it was under this sort of duress that Stephanos decided there was little time to fritter on revisions, editing, and hiring an illustrator. The need for the text was urgent, and he'd earned the rest he'd give himself in the days ahead. He believed his work would somehow make everybody's life instantly better.

Since there were no publishers in the modern sense in ancient Athens, Stephanos headed off to the library, where they could at least direct him towards an appropriate contact. There was an off stillness outside on his way into town. The air was unusually fetid. He was not feeling much better himself. His head felt dis-attached, like a balloon strung high above his shoulders and struggling to keep up with his body. His chest was a bit sore and he was getting warmer from the inside out. He had begun sneezing. *Allergies,* he figured.

Off in the distance thick plumes of dark, foreboding smoke rose from different points all around the city. The river he usually got his water from was today full of naked people languishing, many of them lowing, in agony. He'd get a drink elsewhere. He coughed hoarsely and, when covering his mouth, noticed he was developing some kind of a rash. Further up ahead, Stephanos noticed an atypically large pile of bodies beside a deep, wide hole. The gravedigger leaned on his shovel, taking a breather.

"Hey, excuse me, sir," Stephanos said, "but do you know how to get to the library from here?"

ON THE **PROMETHEAN IMAGINATION**

On the Outside Looking In:
Pregnancy in the Age of Uneven Technology

Sarah Grey

"Whereas once the interior space of the woman was unavailable to the scientific gaze and pregnancy was marked by the woman's testimony that she had felt the fetus move, now the woman's own experience of internal fetal movements is relegated to the unvoiced and unwarranted realm of private experience, while the interior space of the woman is available for all to see as part of the technologized state that Anne Balsamo has called 'the public pregnancy.'"[1]

—Susan Squier

In the ultrasound photo my daughter has her father's nose. This much was clear: everyone on Facebook said so. The day I posted that image, hundreds of Facebook "friends"—true friends, casual acquaintances, former coworkers—saw the inside of my uterus, a body part so private it had never before been visible, even to lovers. It's not uncommon these days for a woman to post her ultrasound as her profile photo, her fetus usurping her face as the image that defines her public persona. And through social media we read one another's birth stories, follow our friends' attempts to conceive via IVF, and watch with concern as premature babies struggle for life in high-tech incubators. Here, too, we debate the risks and benefits of birth control methods, of vaccination, of breast milk versus formula. For women with the resources to access it, the Internet provides space to negotiate the relationship between the innate technology of our bodies and the quickly advancing, often problematic medical technologies designed to fix what we see as our bodies' problems. In such spaces the question arises again and again: how much tech is too much? At what point do we cross the line from helping one another to playing God? Are our pregnancies too technologized? We ask so often, in the less deprived reaches of Western culture, whether we rely too much on our inventions to birth our children—yet the tech most central to our reproduction has been with us always.

When women talk about pregnancy, we speak in stories, from our personal experiences or those of others—so I will speak now from my own stories. Mine was an easy pregnancy; a little nausea, some carpal tunnel, but otherwise healthy and happy. We were excited, eager for the adventure of parenthood. There was a cozy, nurturing intimacy to it. I wanted to hug my love, settle in under a pile of blankets, and grow our child. Even amid the physicality of pregnancy, the aching breasts and the nausea and the heat that pooled in the palms of my hands and the soles of my feet, it was fascinating to see my body performing as an astonishingly advanced instrument of technology. Without any real action on my part it generated an entire temporary organ, the placenta, and within it constructed something *as complex as itself*, something whose capabilities still far outstrip our understanding: a human being. My body took a microscopic, raspberry-shaped bundle of stem cells and turned it into an independent, profoundly complicated organism which, as I write this, is downstairs playing with toy ponies and eating waffles. The programming necessary to create such life is now encoded into her body, too.

I'm the anxious, bookish type, never able to stop

overthinking things, so my usual reaction to facing anything unknown is to hit the stacks: to exercise control by gaining as much information as possible. I read *What to Expect When You're Expecting* from cover to cover more than once, keeping it on the back of the toilet so I could see what fruit to compare my fetus to this month. I Googled everything. We took classes on newborn care, learning how to bathe our tiny child, diaper her, care for her withered little stem of belly button until it fell off.

People asked me whether we planned to learn the baby's sex. "Don't you want it to be a surprise?" I scoffed at this logic. Not knowing might prevent us from getting drowned in pink or blue baby clothes, which would be fine, but I wanted information. "I have to live with this person for the next eighteen years, minimum," I told them. "I want to know everything I possibly can."

Technology gave us some of the best moments of the pregnancy: a chance to *know*, to allay our anxiety in the face of the greatest unknown we'd ever experienced. It made our baby real to us before her birth in ways that affected us deeply. It made us see her more clearly and act in her interests. It was a blessing to us. I looked forward to my visits to the midwife: every appointment was a chance to hear the second heart beating deep inside me. The midwife kept the fetal monitor clipped to her waist like a holster. She would spread a pool of cold blue gel on my swelling belly and slide the metal monitor around in it: *whooshwhooshwhoosh*, impossibly fast, like a little bird. Then a flood of relief when all sounded well inside my womb. The ultrasounds—we had two, the current standard for a low-risk pregnancy—were magical. At twenty weeks the technician pushed hard on my belly, trying to make the fetus wake up and roll over so we could determine the sex. Our uncooperative child, still free for one more moment from gendered expectations, waved a hand instead—middle finger fully extended. We cheered her—it was, we learned a moment later, a girl.

Tech was crucial to the delivery, too, easing my pain and nausea and allowing us to monitor the baby's heart rate, which became dangerously low. The midwife warned Joe that we might need an emergency C-section if this next push didn't do it—but it did. Little Lucia emerged and was placed in my arms bloody and piping hot. We were grateful for the machinery, cold and unfeeling though it felt. (The experience made me skeptical of the trendy natural home birth I had once envisioned: what if we hadn't known she was in danger?) In the year that followed we came to appreciate this even more when a friend gave birth prematurely at just twenty-six weeks. Little Nina, two pounds and change, lived in an incubator for weeks; mother and daughter recovered beautifully. Other friends looked to technical solutions to conceive, undergoing expensive and taxing in vitro fertilization procedures. These sometimes worked and other times left them bruised and grieving.

Unlike some parents we knew, we didn't spring for a state-of-the-art "4D" ultrasound. A misnomer: the picture is still two-dimensional but the technology is more advanced, offering an astonishingly clear view of the developing fetus. The picture is so clear, in fact, that the fetus's features, not yet fully formed, look alien. The photos, inevitably posted on Facebook and greeted with long strings of "Soooo cuuuute!" comments, fall squarely into the uncanny valley.[2] That's not why we didn't do it, of course. We just didn't have the cash to spare; 4D ultrasounds aren't cheap.

As I write this, scientists have just announced the new availability of 3D *printed* ultrasounds. Utilizing the increasingly popular 3D printer and data from the ultrasound, these newfangled contraptions can actually print a little plastic fetus as a keepsake. (There is no diagnostic value to it yet, though some medical schools are printing organs for students to use in practicing for surgery.) News videos showed a young Japanese couple displaying a printed face taken at an eight-month ultrasound next to the actual baby, a few months old.[3] Uncanny indeed.

Is this too much? When I discussed this news story with others, reactions were uniformly negative: "Oh, how creepy!" "Why would anyone want that?" That the procedure's developers had originally used MRI machines, exposing the developing fetus to a medically unnecessary dose of radiation, seemed to underline the madness of the whole endeavor. "It's out of hand," one pregnant friend complained. "There's so much testing, so many machines, and they want us to pay for even more. It feels so cold and impersonal."

The white plastic printed fetus, curled in a uterus-shaped column of clear plastic resin, made me remember the zealots who faithfully protested the abortion clinic where I worked in college. They specialized in decorating downtown Pittsburgh with grisly eight-foot-high signs showing doctored photos of dismembered aborted fetuses, but little plastic fetuses were always to be found in abundance among their number, worn on chains around the neck or scattered along the sidewalk on the legal boundary our volunteer patient escorts strictly enforced.[4] We had them inside, too, in small plastic models of the female reproductive system we kept inside the patient counseling rooms. Many of our patients had been denied basic sex education and knew little about how their own bodies worked; it was left to us to explain ovulation, implantation, and the workings of various birth control methods to women already encumbered with an unwanted pregnancy.

The women at our clinic came from all backgrounds, all ages and races and linguistic groups and walks of life. Most, but not all, were working class. On some we performed ultrasounds if the pregnancy was earlier or later than the usual range. We offered them a choice: they could view the ultrasound or not. In early pregnancy there is little to see, little the eye can understand. Many chose not to, but perhaps half looked. They were curious; they wanted to know. They didn't want the pregnancy but they wanted to make sense of it, to gather information about it to digest later. At eighteen, holding their hands and handing tools to the doctor, I was always moved by this moment of choice. I wondered what I would do in their shoes. Years later, when it was my turn, I looked.

In 2014, thanks to the tireless machinations of anti-abortion politicians, ten US states now require women seeking abortions to submit to an ultrasound first, whether it is medically necessary or not. The ultrasound procedure in early pregnancy, when the vast majority of abortions are performed, is transvaginal: a hard plastic wand is inserted into the woman's vagina and pressure applied to her full bladder. Rape survivors are not permitted to decline this violation. In three states, women are forced to view the image even if they do not want to see it; in seven, doctors are required to offer them the option of viewing it.[5] The cost of buying additional ultrasound machines and hiring technicians to operate them drives up abortion costs, making it more difficult for women to access the procedure and driving some clinics out of business—which is, of course, the ultimate goal of such laws.

What's particularly astonishing about this invasion of technology into the most intimate human spaces is that such tech is often unavailable to women who desperately need it. The cost of an ultrasound in the United States can run more than a thousand dollars,[6] and even pregnant women who have insurance coverage face high deductibles and out-of-pocket costs for every technological intervention, from simple ultrasounds to the complicated fetal heart scans or amniocentesis used to monitor high-risk pregnancies. (Independent "crisis pregnancy centers" advertising free ultrasounds in cities are almost always run by anti-abortion religious groups; a congressional investigation found that eighty-seven percent of such centers provided "false or misleading information."[7]) Premature births and low-birthweight infants, too, require advanced life-saving technology—often at astronomical expense. And since poverty is strongly correlated with late or no prenatal care, which puts women at far greater risk of premature birth and low birthweight, enormous medical bills often end up in the hands of the families least able to pay them.

The picture is even grimmer in developing countries: in Uganda, an astonishing 310,000 women die every year from complications during pregnancy and childbirth; of every thousand live births, sixty-three infants do not live to see their first birthdays.[8] Royal Phillips, a Dutch corporation, launched a project to provide eleven rural Ugandan health centers with ultrasound technology; doctors there credit its use with early intervention that is saving thousands of lives.

The ultrasounds not only help medical staff predict and plan for risks, they also attract women who go on to receive additional crucial prenatal care, including malaria and HIV testing. The centers cost about ten thousand dollars each—the cost of twenty 3D-printed keepsake fetuses.

Perhaps the problem, then, is not that technology infringes too much on the sacred intimacy of pregnancy. Perhaps the problem is simply that technology is distributed unevenly. Middle- and upper-class women in developed countries can take their access to lifesaving medical technology for granted, but have unwanted technological procedures pushed on them because they are seen as, above all, a highly profitable market. Women who attempt to terminate an unwanted pregnancy have this tech forced into their bodies against their will, against all medical necessity, in a cynical attempt to make termination inaccessible. And in countries and poor communities where it is desperately needed (and yet entirely unprofitable), women die by the hundreds of thousands because of its absence.

There is much we still don't understand about fetal development. Sometimes there is a genuine gap in our knowledge; other times there is simply no funding for research, particularly when findings about, say, environmental pollution or food additives might endanger corporate profits. Pregnant women, whose bodies are the ultimate 3D printers, deserve as much information as they want and need about their pregnancies. For some, this means shelling out a wad of disposable cash for a ghostly plastic model of a still-forming face; for others, it might mean walking a hundred miles on swollen feet just to access basic prenatal care.

It is understandable that a generation raised amid the

most rapid technological revolution the world has ever seen is asking difficult questions about how much technology we can incorporate into our wombs and our lives without sacrificing something of our humanity. Are we less human if we become more cyborg? In the final analysis, though, the true test of our humanity isn't how much technology we choose to embrace; it's how many of us actually have a choice.

Piddil Iddil Iddil

Sara Graybeal

puh, puh, puh puh puh
puh, puh puh POOM
puh, puh, puh puh puh
puh puh puh puh POOM

My students have discovered Google Translate.

On Day 2 of their after-school program, barred from
Call of Duty, Grand Theft Auto V, and twerking videos,
they click restlessly on desktop icons: Solitaire, Clip Art,
in desperation, Recycling Bin. They design logos in Paint—
free us, me, them, you, Ronisha spray paints across the page—
and make PowerPoint animations of the car races they are
not allowed to play. When these options fall short,
they ask Google all the questions in the world:

wat happen when u stuff a cat with cotton balls

wen ur pregnant do u really eat dirt

They do not wait for answers. The point is to ask,
to engrave one's words on the screen. But even this
grows old, moods worsen, and though they still insist
on computer time each day, they are no longer sure why.
I suggest Google Earth—"You can find your block.
You can go anywhere in the world"—but we are afforded
one glimpse of our blue planet before the monitors freeze.

"Yo, it's the government," says Craig. I start to say, "No, it's the technology," but this is not different enough from his point to justify rephrasing it.

Then Zahir types *trans* into Google. He does not know what this means but he thinks it will be interesting. The first link yields two boxes, eighty languages, and a microphone icon in the corner. He says *Craig loves Amani* in Haitian Creole. He says *Malcolm wet his pants* in Slovak. The mechanized woman talks for hours and never grows hoarse:

Craig renmen Amani

Malcolm mokré nohavice

The next day, the room fills with hisses and pops. Rhythms shift in each language; *ppppppp* bursts into a thousand sounds like a dandelion blown to seed. They align their computers and play them in unison:

p p p p ppppppppp

I went to school, got suspended. Drove to the store and got rear-ended.

p p p p pppppppp

I got bars, bars for days. Ant can't write no bars my way!

They layer Norwegian on Zulu, Punjabi on Basque. They forget to do their homework and they forget to misbehave. There are rap battles and sing-offs and poetry and there is Cristina, who has learned *the best grandma in the world* in Spanish and can't wait to get home and try it out. This is a planet as huge and lovely as the one that froze their screens, and it is their own, their birthright, as I was born to cassettes and VCRs, believing home videos were the best way to have fun. My students are the smiths of these computers, twisting what is before them into new, beautiful things.

puh, puh, puh puh puh
puh, puh puh piddil
puh, puh, puh puh puh
piddil iddil iddil iddil iddil

When my students leave, the computers flicker off, knowing that we have nothing to say to each other. They will be rested and ready tomorrow at three o'clock, when their screensmiths walk back through the door.

EXCERPT FROM THE FORTHCOMING NOVEL *AGE OF UNCERTAINTY*

Nicolas Esposito

"Okay everyone, let's try this once more. Can anyone in this room give me two examples of chemical compounds that are considered isomers?"

The auditorium went silent. The two, maybe three people who were possibly taking notes on their laptops now ceased the tapping of their keys. The steady drone of whispering in the back of the room quieted. If a janitor had mistakenly entered the room at that moment, he would have started emptying the trash or sweeping the floor with a complete lack of self-consciousness.

The professor took this opportunity to turn back to the whiteboard, let out a long sigh, and tuck in the shirttail that had billowed out as he paced across the stage. He ran his left hand across his head, stretching his fingers over the large swath of bare skin to catch the patches of hair that lined his ears. He reached behind his ear and rested his hand on the device he had installed a few months back to improve his dwindling hearing. Rather than find the volume dial, he pushed his ear forward, as if to try and catch an echo. But there was nothing.

"Yes, yes, that's correct, cyanic and fulminic acids. Congratulations to the person who blurted that out. But please remember, let us raise our hands. One at a time, please."

He paced the length of the stage again and continued. "As we read in our books, the term *isomer* is translated from Greek as meaning *equal parts*. Meaning that each compound shares an equal number of elements, yet is rearranged in different sequences to create different compounds. Now, can anyone tell me which elements make up these two compounds?"

He took a pause and scanned the front half of the stadium seating, which contained about one-third of the one hundred and fifty students the registrar had packed into his course. He settled on a young woman in the front. "Yes, you, um, you, in the pink sweatpants."

"But I didn't raise my hand," she quickly countered.

"Oh, you didn't, yes, I see. But how about a stab at it anyway."

The young woman shifted in her seat and pulled on her knees as if her legs were asleep. "Um, well, I would say that the compounds are, most likely, oxygen."

The professor's eyes lit up, and he started to nod his head with his chin leading the way.

"Um, nitrogen..."

"Yes, yes, keep going," he began to breathe heavily.

"Heel..."

He clenched his teeth and quickly shook his head from side to side.

"Hydrogen," she corrected herself.

"Yes, yes, and one more."

The girl looked at her notebook, then over to the student sitting next to her, then back to the notebook. "Chlorine?"

"Ah, you looked at your notes and you saw that 'C,' I understand. But without the 'l' it's not chlorine it's..." He trailed off, waving his fingers in a circular motion.

"Cadmium."

"*Carbon*, yes, that's correct. The most essential element

on this planet. The stuff that makes the soil and the trees and our bones. The building block of life."

He walked to the edge of the stage. He loved this effect that worked so well in his first few years of teaching. The closer he got, the students would respond by lifting their backs off their chairs and leaning into his lecture.

"Isomers were one of the most substantial discoveries of what is now modern chemistry. It proved to scientists that compounds were not only products of the natural environment, but that chemists themselves, in laboratories, could actually combine the known elements of the periodic table into synthetic compounds. The discovery of isomers was so important, so life-changing because it unlocked these codes. It allowed scientists to artificially sequence elements into compounds that were once only found in nature. It was the discovery that first gave the world the synthetic version of Urea."

"Urea, like pee," a disgusted voice echoed in the silence.

"Yes, yes, like pee," he exclaimed dramatically, his toes hanging off of the edge of the stage.

He didn't want to lose this momentum so he built off of the last comment. "So why is this discovery important? Why was synthesizing urea such an achievement? Young lady in the pink sweatpants?"

"I didn't say anything," she shouted out.

"Right, right," he reached for the volume dial in his ear, but couldn't get his fingers around it, "you've done your part. Um, how about you, young man with the football jersey."

The professor took a risk on this student who sat a few rows back. He knew football jerseys were like badge credentials that alleviated these young men from ever having to speak in class. He brought his attention back to the first few rows and focused in on a beam of light reflecting off of one young man's face.

"Yes, you with the metal bars through your chin and eyebrow. Why was the discovery of urea so important?"

"Huh," the young man said, as he pushed himself up in his seat and sat forward.

"Urea," the professor prompted, "why was it so important that scientists were able to synthesize this compound from their knowledge of isomers?"

"Um, well..."

"Yes, yes," the professor urged him on, this time with almost half of his foot hanging off the stage.

"Well, it's-because..."

"Uh huh!" the professor shouted.

"I-don't-know."

The professor slowly inched back from the precipice of the platform. He took one step back, then another, and slowly walked behind the podium and put his head down for a moment to collect his thoughts. The muffled static of his breathing was all that filled the space, until the shuffling of papers and the zipping of book bags roused him once more.

Still looking down, he spoke directly into the microphone attached to his lapel, "We still have twenty minutes left."

The shuffling stopped. The zippers halted mid-zip. And the professor began again. He turned his back to the students and ran his fingers through his few whiskers of hair.

"We've been going about this all wrong," he said.

"Excuse me, what was that, I didn't get that note," one studious but non-contributing young woman asked from the middle of the room. The professor turned back around and let his voice project into the auditorium, "We've been going about this all wrong," he boomed. "You all are from the generation of personality, of people. We should not be dealing in concepts or facts. We should be dealing with the people who created these concepts and who uncovered these facts. Here, I'll show you, an easy example that has nothing to do with science. Who here can tell me the tactical reasons for where the first battle of the Revolutionary War was fought?"

He didn't even wait for an answer that he knew was not coming.

"You see? Now, who can tell me one of the major generals who fought in the Revolutionary War?"

Again, silence.

"Okay, okay, maybe that was a little before your time." He sighed in exhausted frustration. "Something more modern. When did we first land a person on the moon?"

Silence.

"Okay, now, who was the first person to step foot on the moon?"

"Buzz Armstrong," a voice confidently shouted out.

"We're getting closer, we're getting closer. One more try. Where did Apple computers get the inspiration for their desktop and interface?"

A few murmurs, but nothing coherent.

"Now, who founded Apple computers?"

"Steve Jobs," a chorus of voices shouted out.

"Yes, yes, Steve Jobs. You see, the person is always more recognizable. People are always more interesting. And in the case of isomers, this is no different. You see, isomers were discovered by two scientists in the mid-eighteenth century, Frederich Wohler and Justus von Liebig. I'm not even going to ask if any of you know these names because some people in the scientific community may not even know them. But what if I was to tell you that it was once said that Justus von Liebig was to chemistry what Sir Isaac Newton was to physics."

Suddenly, the unexpectedly engaged young man with the piercings chimed in, "Yeah, Newton, he invented gravity."

"You know, I'm going to go with that. Yes, invention, discovery, what's the difference? And what if I was to tell you that von Liebig's inventions make up a majority of the advancements that make us modern. Synthetic potassium cyanide that gave way to the photograph, synthetic fertilizers that grow your food, chloroform that led to the advances in modern surgery. Of course, he didn't invent any of these things, just like Isaac Newton didn't invent gravity. But isn't discovery just as equal?"

No one answered this question, but he could feel the students following, so he continued, "I must posthumously apologize to Mr. Wohler for not focusing on him, but Justus von Liebig is just too interesting. As a boy, when asked in front of his primary school class what he wanted to be when he grew up, this audacious little scamp replied, 'A chemist.' I'm not sure how many of you want to go into this field today, but in the early nineteenth century this was laughable, and that's what his peers and his teachers did. They laughed at a young boy who wanted to go into a field that was looked at as nothing more than an accessory to pharmacy. But I assure you, by the end of his career they were laughing no more.

"Although his father did not quite understand his son's obsession, he didn't laugh at it. He sent him to work in the apothecary of a local druggist. But von Liebig's keen curiosity got the best of him. As with the tale of curiosity and the cat, von Liebig would sneak away from the boring task of pill-making to go up to the apothecary attic and conduct his own experiments. It was not long before he made the wrong mixture and completely blew out the windows of the attic, almost killing himself in the process."

A light chuckle rippled through the auditorium. He had them, so he tightened his grip. "As the young man grew into adulthood, his boldness did not cease. His ambition and talent secured him a seat at the University of Bonn to study with the great chemist Dr. Kastner, who was bringing respect to the field. He excelled, but would debate and compete, and even quarrel with his fellow classmates over theories and concepts. There was talk that he was arrested under suspicion of political offenses."

"But he was not always disagreeable. Von Liebig could also be quite magnetic. The young poet Platen, who met von Liebig at University, described him as, 'never more beautiful, with brown eyes and a slender form that attracted one instantly.' The young poet said of this relationship, 'after so many deceptions,' that he had never found more peace and happiness in a friendship, and that it was love at first sight."

A few *oohs* and cackles sounded off, and the professor responded, "Yes, yes, take that for what you will. But my point is that the man was enigmatic," he said as he began to almost skip from side to side of the stage.

"This personality and ambition earned him the degree of Doctor of Philosophy at the age of nineteen. It was around this time that he published his first paper on the composition of fulminating mercury. Now what does this word bring to mind, something we were talking about earlier?"

"Fulminic acid!" the young lady in the pink sweatpants shouted.

"Yes, yes, well done."

The students were actually connecting the dots on the points of his talk. He felt like he was an associate professor again.

"It was this paper that led to the next important moment in von Liebig's life. His research had progressed into fulminating silver, and after delivering a paper on the subject at the Academy of Sciences, he was approached by a man while packing up his things. The man was intrigued by von Liebig's ideas and invited him to dinner. But von Liebig, in all of his kinetic energy, lost the address. The next day, a friend asked him how he could be so brash as to not show up to a dinner at the request of Humboldt the Traveler and Dr. Gay-Lussac. Liebig described his reaction as nothing less than 'thunderstruck.' He rushed to Humboldt's estate, and was eventually invited to dinner."

The professor paused, wondering if they were following, if they even cared. He didn't hear any zipping or whispering, so he took it as a sign to proceed.

"I don't expect you all to remember these names, nor will I quiz you on them. But take this as a life lesson. For this meeting was the defining moment of von Liebig's life. It was due to these people that he was given a professor's chair at the small University of Giessen. Although not the most prominent university in Germany, it was where von Liebig would become a legend."

"It was here where he first experimented with the actions of mixing chlorine and alcohol, which gave man chloroform, and consequently, the beginnings of modern surgery—a patient could now receive anesthetics during a procedure. It was here where his extensive agricultural experimenting on test plots of farmland revealed to him that plants actually received their elemental nutrition such as carbon, oxygen, and nitrogen through the air. And it was that discovery of nitrogen's return to the soil through the release of ammonia gas from decaying organic matter, whether it be manure or leguminous vegetables, that led to von Liebig's synthetic production of nitrogen for plant growth—an advancement that gave us industrial farming, not to mention the explosives industry. Sure, von Liebig tested these hypotheses in the laboratory, but he conducted his research literally in the field, working the land like a farmer.

"Von Liebig was the first person to create potassium cyanide from melting two different types of potash together. Potassium cyanide was then found to dissolve many silver salts, which was crucial for early photographers to remove silver compounds from their negatives. And it was von Liebig's sometimes vicious competition with Louis Pasteur that fueled the refinement of fermentation and pasteurization that allowed for the mass production and preservation of food."

He looked at his watch. The minute hand ticked into place on the forty-nine as the second hand began its final revolution before the end of the period. He needed to bring it home.

"Von Liebig was the ultimate collaborator and teacher. Although the Nobel Prize was created some years after von Liebig's death, his Giessen laboratory would go on to produce dozens of prize winners in chemistry. But the most important partnership he ever had was with Frederich

Wohler. While von Liebig was predisposed to rash levels of outspokenness, even managing to almost lose his hearing and sight when a mixture of fulminating silver and ammonium sulfide exploded under his nose in his later years, Wohler's calm demeanor and tempered opinions established a balance. But it has been said that both men's commitment and dedication to seeking out truth was what bound them. It was this fact that made them two of the most compelling and important figures in modern chemistry."

The words "modern" and "chemistry" erupted from his throat. He was once again on the edge of the stage. He had no more to say. It was up to the students to take this information into the next epoch of chemical discovery. He looked into the auditorium as far as his eyes would allow, and at that moment, without even the ceremoniousness of a bell, the synchronization of the students' computing devices prompted them to stand up and walk out. His fifty minutes were up.

He watched as the students filed out the door, in no apparent rush to their next fifty-minute session—if there was one. The professor turned around and tucked in the shirttail that had come loose again. He stretched his fingers through those bits of hair close to his ear and adjusted the dial on his earpiece. Apparently, he had forgotten to turn it on.

MARS: THE PIONEER

SCRIPT FOR AN ISSUE OF A GRAPHIC NOVEL

Tom Coiner

Illustrations by Kristy Caldwell and Ray Jones

where a new paragraph indicates a new panel
and pages are clearly marked
while speech is indicated in bold
and sound effects abbreviated to SFX

Cast of Characters (In order of appearance)
Doctor Marvin Geist
Odetta Haven
Captain Jared Crocker
Doctor Naomi Woods
Lieutenant Benjamin Hensley

PAGE ONE

A vast desert fills the page. The curve of the horizon, ripples of sand dune tops. A figure, its shadow stretched by a low sun, moves forward across the sand. Figure, shadow, and tracks are all barely visible. All also very, very red.

PAGE TWO

The figure is clearer now a hundred yards away. Gritty wind blows. Two faint oblong moons hang in the sky.

PAGE THREE

Ten yards. The figure is wearing what looks like a HAZMAT suit.

Torso. Name tag: GEIST.

Face through domed plexi helmet. DR. MARVIN GEIST. Stubbly, disheveled, unsexy. Breath fogs his glasses and helmet. Eyes open a little too wide, lips dry and cracked.

M **So...cold.**

Marvin's armband digitool readout displays "O2 10 MIN REMAIN."

M **Hunh...recipe for remains...**

PAGE FOUR

Marvin stops. He breathes. The view is awesome.

His hand slaps the armband digitool.

It reads COMMS DOWN.

We see what he sees. Nothing ahead nothing behind.

SFX **Bzzt!**

Voice **Dr. Geist?**

Tears shine in Marvin's eyes.

Voice	**Marvin?**

PAGE FIVE

A child at a drafting table hard at work on the previous image of Marvin crying. August 5, 2012, Appalachia. Charcoal and paper.

Mom	**Marvin!**

Young Marvin has drawn all of the previous four pages' images. He is in Goodwill clothes out of the '00s and glasses out of the '60s.

YM	**Wha-at?**

Young Marvin's room is a disaster. Comic books, laundry, a mock spacesuit, paper covering walls and floor.

YM	**Uh...**
Mom	**Coming up now...**

PAGE SIX

He whirls. Edgar Rice Burroughs.

He grabs. Roald Dahl.

He stuffs. Ray Bradbury.

He conceals. William Gibson.

PAGE SEVEN

Mom opens door to find Marvin clutching a stack of comic books. She is in beat-up overalls and paper-thin flannel.

Mom **Ok. No television.**

Horror!!!

YM	But Mom, Curiosity!
Mom	I'm sure if the Curiosity rover was here it would tell you to clean your room and rake the backyard...
YM	...it couldn't fit in here...

Marvin pouts and schemes.

YM	Mom, I'm not sure you understand. For seven whole minutes the lander will go out of communication and that whole time it could blow up in the atmosphere!

Mom in the doorway but she is not the one talking.

Not Mom	Congratulations, Dr. Geist. We made it through minute seven.

Young Marvin's face is confused.

Not Mom	Dr. Geist...

PAGE EIGHT

ODETTA HAVEN, 37, whispers/ glares at older Marvin of the unsexy stubble from before. The glasses are the same he wore as a child.

OH **...it's your turn.**

A MAN standing at a podium in a small private lecture hall beckons Marvin.

MAN **Care to say a few words, Dr. Geist?**

Marvin, bed head scruffy in a white lab coat, surprised to find himself here now.

PAGE NINE

Marvin stands, fist raised to mouth, clearing throat.

M **Uh, yes.**
Kahem.

Marvin lays his notes on the podium.

M **Thank you, Paul.**

Marvin at the podium.

M **Ladies and Gentlemen...**
Uh...

PAGES TEN AND ELEVEN

Marvin clicking on the overhead projector stands before a giant image of the Red Planet against the blackness of space. Our title page.

M Mars!

PAGE TWELVE

Outside the Haven Science Center, University of Kentucky, Lexington, 2035.

Inside the Science Center. Marvin at the podium.

Marvin's legal pad covered in doodles.

M Thank you all for coming.

We see coffee-drinking faces and yawning faces. Some of them are famous faces.

M We lied on our invitations. This is no TED talk. This is exclusive content, not being filmed or recorded in any way.

We see skeptical, bored, texting faces. Some of them are rich, powerful, famous faces.

M A one-time performance for your benefit. My name is Dr. Marvin Geist...

We see faces too rich and powerful to be famous. They listen.

M ...and I'm the lead engineer at Geist Propulsion, a subset of Haven Corp. Mars is my project.

PAGE THIRTEEN

Marvin has one arm reached behind him clicking the power point.

M As you all know, man first walked on the moon in 1969.

Marvin centered at the podium.

M What you may not know is the last time anyone walked on the moon was 1972.

The audience viewed from behind Marvin.

M If you'll forgive me the history lesson...

PAGE FOURTEEN

Marvin smirks.

M ...In 1905, soldiers were riding *these* into battle.

SFX se-Click!

Antique sepia photograph of a war horse in a field, posing proudly.

Marvin again.

M — Forty years later we invented this...

SFX — se-Click!

A nuclear bomb explodes in the ocean.

PAGE FIFTEEN

The gathered assembly unimpressed.

PAGE SIXTEEN

Neil Armstrong stands with his aluminum flag.

M — Twenty years after that, we were on the moon. Since then? Well, we've landed a rover or two.

Curiosity on Mars.

M — But no man has set foot on a celestial body since.

Marvin holds up his iPhone X.

M — You all carry in your pockets more computing power than all of NASA had in 1968.

Marvin clears his throat.

M — Ka-hem.

Excuse me, but what the FUCK!?!

PAGE SEVENTEEN

Crowd faces uncertain.

M — Sorry to say I've got news for you.

Marvin.

M — It's called the human *race*.

Some faces uncomfortable, some swaying.

M — Every year we make microscopic changes to the same half-broken pocket tech and we create the illusion of moving forward when in fact we are spending our tremendous resources on *slowing down*.

Marvin's lecture notes sit on the podium, his speech slows, triple underlined.

M — After slow comes stopped.

PAGE EIGHTEEN

The crowd uncertain.

M — It's not news—we've put Earth on its deathbed.

Marvin stands tall.

M — Which is why I have called you here. Entrepreneurs all, I offer you an opportunity, to take part in a revolution. From dead land...

SFX — se-Click!

Barren Mars.

SFX — se-Click!

Mars and Earth next to each other,

Earth giving Mars pieces of itself and Mars giving back itself in return.

M — Terraforming. From the Latin terra, meaning Earth...and form, as in transform.

PAGE NINETEEN

Marvin.

M — We are at a critical point where our own inventions threaten our existence. Curiosity may kill the cat after all. But in truth, a missile is simply a rocket that lands. I propose we tilt our rockets again to the sky, and never ask them to come back.

SFX — se-Click!

A schematic appears on screen behind Marvin, out of focus.

M — First comes water. Without which there is nothing.

PAGE TWENTY

Focused in, they are oil industry schematics.

M — We'll have to dig for it, and, Curiosity said we'll find it.

Computer simulation of ice caps melting.

M — Then we capture Mars's own green-house gases to heat the planet up, melting its ice caps, and forming a Northern sea.

Mars with water covering most of the Northern Hemisphere.

M Filled with water, hungry for life.

SFX Bzzt!

M Everything we need up there we have down here in spades. So why not?

PAGE TWENTY-ONE

Marvin continues, but Fred is seen now way in the back, classic chain-smoking stage manager, communicating with Andy, the satellite uplink man.

M The water irrigating our terraforming fields will be purified...

F Negative, Andy. He's up there talking now.

Marvin continues.

M ...and recycled to sustain our crew.

Fred on the headset, now we can see he's at a soundboard in the back of the auditorium.

M Nothing will be wasted in the wasteland of Mars.

F Well, get them back *behind* schedule...

We're with Fred, Marvin talking in background.

M My thanks to Ms. Odetta Haven, who footed the bill for this project...

F Let me talk to 'im.

Odetta Haven smiling at the acknowledgment.

M ...the public sector being the limpest noodle in the pool...

Back to Fred while the crowd chuckles scattered *hehs* and *hahs*.

F Cut the feed, Andy, fuck the fourteen *minutes*!

PAGE TWENTY-TWO

Marvin gestures down at Odetta, who is ready to take the stage.

M Under Odetta's father, Haven Corp. blew up the surfaces of Appalachian Mountains for coal. Now it's running giveback programs across West Virginia and Tennessee. Under Ms. Haven's leadership, they've pioneered new technologies designed to restore Mountaintop Removal Mining sites back to their pristine nature.

Odetta in Appalachia doing a hard hat photo op, environmentalists on the right, senators on the left.

M — **Haven's tech is the basis for our terraforming. The parts, labor, and brains behind this operation are 100% Appalachian...100% American. Because we're gonna drag you kicking and screaming into space if it's the last thing we do.**

Odetta glares up at Marvin with hooded eyes.

M — **Ahem...stimulating employment of 7,000 West Virginians to start. So, without further ado...**

SFX — Bzz-cha-cha-vrmmmmmmmm

PAGE TWENTY-THREE

Crowd looks up as a disembodied voice blares over the

speakers. JARED CROCKER is an all-American Captain America, confident master of all he surveys. We'll meet him later.

JC	**Do you have me Barsoom? Transmitting to Lexington?**

Marvin covers his surprise. BENJAMIN HENSLEY's voice responds over the speakers; he is Jared's senior in years but junior in rank, the crew's astrophysicist and technical engineer.

BH	**That's an affirmative. We are bridging the comm. Signal clear.**

Marvin scrambles.

M	**Uh, ladies and gentlemen...**

The crowd murmurs.

BH	**Lo-res packets should be delivered momentarily.**

PAGE TWENTY-FOUR

Marvin louder.

M	**If I may—**

Odetta Haven back in her seat. She seems pleased.

Marvin into an earpiece.

M	**(Fred! Can we record and play this later?)**

The projector screen erupts in moving static.

SFX —bzzrt—

PAGE TWENTY-FIVE

On the screen a giant lo-res interior of a spacecraft appears.

Marvin cranes his neck to see behind him.

JC	**The packet contains first series ten photos. And now, open the pod bay doors, Hal.**

The cockpit of Barsoom comes into focus.

JC	**Suits on, ready to lose it.**

PAGE TWENTY-SIX

Marvin clutches for control.

M	**Ladies and gentlemen!**
BH	**Depressurizing in three—**

Lo-res image of the Red Planet's surface through porthole of Barsoom.

Marvin alone.

M	**God...uh...if I may present...**
BH	**Two—**

Closer on Marvin.

M The head of the crew of the Barsoom.

BH One.

PAGE TWENTY-SEVEN

A lo-res shot of the Captain as he steps into depressurization chamber.

M Captain Jared Crocker.

SFX Hisssssssssssss

Marvin speeds to the finish!

M We wanted to ask for your help in going to Mars. But we thought we'd have a better chance of getting it if we had...already arrived.

Jared is still on the screen.

JC Stepping out.

PAGE TWENTY-EIGHT

The audience stares up at a lo-res image of Benjamin Hensley at the comm board inside the cockpit of Barsoom I. The Barsoom is a permanent habitation module; it has a cockpit because it was until very recently a small part of the Phobos I Mars Rocket, recently landed, crew of four. The HAB will billet the astronauts until their blastoff and journey home aboard the Phobos II Earth Return Vehicle.

M	These images are the first seen by human eyes on the Red Planet. They were taken by Benjamin Hensley—
NW	Are the packets coming through, Barsoom?

Marvin has heard something no one else heard.

M	Naomi!?

PAGE TWENTY-NINE

Flashback: young man and young woman leaving church, flower petals in the air.

Marvin's left hand bears a ring tan line.

Marvin pales.

M Er, no, that's not right–

PAGE THIRTY

Marvin has frozen.

Side view silhouette of Marvin above the masses.

M Ka-hem.

BH Good shooting, Dr. Woods.

Marvin sweats.

M Correction, er, the photos are being taken live by my, by...NASA, our crew's medical doctor and biologist, er...Dr. Naomi Woods.

Lo-res selfie; Naomi Woods, full EVA. Crew's medical doctor, astrobiologist, Marvin's ex-wife.

JC And...boots on red dirt!

Marvin is trapped.

M You're hearing Jared's voice with a delay of fourteen minutes.

PAGE THIRTY-ONE

A silhouette: the audience from Fred's perspective.

NW	It's so...cold.
M	Naomi's lo-res photos are arriving slightly later.

Lo-res shot of Jared descending EVA ladder outside HAB.

JC	One small step of Mankind is one giant leap on Mars.

Marvin frowns.

M	I think he means to say–
JC	Whoooooooooo–
M	In fact Mars has more than two times the gravity of the moon.

The audience smiles up at the screen.

JC	–hooooooooooo!

Marvin grumbles.

M	Ridiculous.
JC	I fucking flew!!

SFX Jared huffs.

PAGE THIRTY-TWO

A low-res pic the Barsoom hatch door.

M	This is never before heard audio, live-piped directly here...without editing.

Jared and Naomi EVA looking back through the porthole.

M	The first woman and man on Mars.
JC	Proceeding towards ISPP Rover.

PAGE THIRTY-THREE

A computer simulation of launch and landing. The rover and ERV are shiny and new, Mars flat, welcoming, barren, lonely.

M	Six months ago...

The Earth Return Vehicle lands.

M	An Earth Return Vehicle, or ERV, landed on Mars and deployed a rover, similar to Curiosity...

ISPP Rover leaves the ERV.

M	...its job was to convert Martian air into rocket fuel.

ISPP deploys antennae.

JC	Are you seeing this, Barsoom?

3D molecules interact.

M — Though that sounds preposterous, Mars's atmosphere is 95% CO_2. By combining this CO_2 with hydrogen brought from Earth, the ISPP unit produces methane and water, and...

NW — Check out these rocks.

JC — Wow.

PAGE THIRTY-FOUR

ERV blasts off from Mars.

M — ...and,the methane-oxygen rocket fuel generated in the six months the ERV has been on Mars is enough to get Captain Crocker...

NW — Jesus.

ERV parachutes into the Atlantic.

M — ...and his team blasted, uhh...all the way back to splashdown on Earth.

Wide shot of the audience. Jared's line intrudes from the upper left, urgent.

JC — Barsoom! Are you seeing this?

BH Somewhat. Signal is muddy.

NW Jesus Christ!

PAGE THIRTY-FIVE

Marvin screams into his earpiece.

M What is this goddamn screaming? Fred, visual.

Fred pushes a million buttons.

Lo-res cluster of objects, blurry at a distance.

M Is that the most recent pic?

SFX se-Click!

PAGE THIRTY-SIX

A circle of red rock spires surround and tower over the rover. It is barely visible inside its ring of rocks. Jared Crocker stares up at the rocks, Marvin stares up at Nate on the screen along with the audience.

JC How'd the rover get in there?

M ...curious geologic occurrences... common on Mars.

JC I'm going in.

PAGE THIRTY-SEVEN

A gap between spires big enough for a human.

JC	Rover hydrogen reserves are full. Methane and oxygen produced is about 1.7 times what we expected.
NW	Upload its camera data.
JC	Affirmative.

Confused, concerned, excited faces in the crowd.

M	Uh...ladies and, this is the first step on the road to human habitation of Mars.

Marvin.

M	We can make this journey again and again, with your generous—

SFX Kkkkkkkkkkkkkknnnnnnnnnnnngggggg!

PAGE THIRTY-EIGHT

Again the screen transfixes everyone where a lone red rock juts up.

JC	Movement, Barsoom, movement!

SFX Hissssssssss

BH	Crocker, switch your vidcam on.

Everyone listens.

JC **Oh my God.**

SFX Hisssssssssssssssssssss

Marvin holds his breath.

JC **...movement...can't...quite...see it...**

Silence.

The screen.

PAGE THIRTY-NINE

The whole lecture hall.

NW **Jared! Don't!**

SFX KKKKKrrrrrrunchhhhhhhh!
SFX HISSSSSSSSSSSSSSSSSS

NW **No, no, nonono! GET AWAY FROM ME!!**

SFX Crack.

PAGE FORTY

One crowd face is blank.

Another afraid.

A smartphone slipping from fingers.

PAGE FORTY-ONE

Odetta is smiling unnoticed, her face illuminated from below. Odetta texts on her phone.

TEXT To: **Adrian**
And the Oscar goes to...

SFX Bzzrt!

Fred's face.

F **Marvin...**

SFX Bzzzzzang!

PAGE FORTY-TWO

The blank crowd face.

BH **Lexington, this is Barsoom.**

The fearful face.

BH **If you're hearing this, the extravehicular team has gone full dark as of 10:37 a.m., Eastern Standard Time.**

Smartphone, cracked screen, on ground.

PAGE FORTY-THREE

Marvin, the screen at his back.

BH Will attempt re-establish and await your instructions. Over and out.

PAGE FORTY-FOUR AND PAGE FORTY-FIVE

Sun sets over the Martian horizon behind Captain Jared Crocker. In one hand he holds a long, heavy wrench bearing the logo *Haven Geist - Mars*. In his other hand is a Pioneer-brand oxygen pack with hoses ripped from their couplings. Below him a figure lies crumpled, helmet cracked and bloody, steam rising from the holes in her suit.

END OF ISSUE

Frontier 1.2, 20"x30", archival print, 2013

Frontier 1.0, 20"x30", archival print, 2013

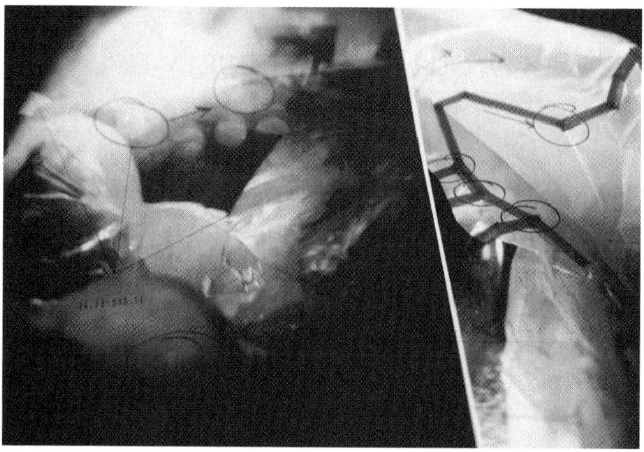

FRONTIER 1.1, 20"x30", ARCHIVAL PRINT, 2013

Kathryn Zazenski

ON SETTING COURSE TRAJECTORY

Big Data Valentine

If person #2's liquid assets > fixed assets, then	.60
Calculate Affection Quotient. # of times A in name #2 / # of times A in name #1	-.60
Calculate Loyalty Quotient. # of times L in name #2 / # of times L in name #1.	.70
If person #2 has no children, then	.50
If person #2 has been married twice before!	-.70
If person #2 likes aphrodisiacs & eats asparagus (retrieve from Shopper Database), then	.20
Bring back #2's results from NIH Blood Sample Database. If serotonin > norepinephrine, then	-.46
If #2 has receptors for oxytocin released during sex / age (obtain from Social Security Database), then	.12
Take from state DMV Driver License Database photos of #1 & #2. Cross match. If distance between eyes =, then	.27
Capture from NSA, NYPD, ATM surveillance. Cross match #1's father's physical features with #2's physical features, if =	.36
Probability of successful relationship = SUM(O18:O26)	.99

BE MINE!

Jacalyn Shelley

Craigslist Time Machine (or, A Missed Connection)

Eric Smith

In the event the wildest dreams of H.G. Wells, Robert A. Heinlein, or Ray Bradbury came to fruition, and time travel suddenly became available, I sure would love to write myself a Missed Connection on Craigslist and tell my past self not to meet my future wife.

Okay. Before we continue, I should explain a few things.

I don't want you thinking this is some silly comedy stand up "take my (future) wife, please!" shtick, or that I'm some sad, miserable husk typing away in front of a dim screen, overwhelmed with regret and rue about choosing the wrong woman. In fact, while I'm writing this, she's lying on the couch, snuggled up in a black and white fleece blanket, her curly hair cushioning her head against an already soft pillow. And she's looking as lovely as ever.

It's because we almost met. Seven years ago. In 2007. Because of Craigslist.

Let's imagine those vehicles dreamed up in classic science fiction do exist, and we hop in Herbert's time machine and travel to the fall of 2012. I was working on an article for a local Philadelphia paper and wanted to share the piece with my now fiancée. It was an article I was particularly proud of, and I wanted to make sure I got it right. But for some bizarre reason, she wasn't getting any of my recent emails with the article attached. After several tries, she asked me what email address I was using. So I told her.

"How do you have that email?" She immediately asked,

surprised and confused. "I haven't used that email address in five, maybe six years!"

After searching in my Gmail, I found a number of our messages to each other, sent over the past few months. Archived moments from the earliest stages of our relationship, many of which persist today. There was nothing terribly surprising. Emails to plan (Groupon) dates, links to nonsense Top 10 lists from around the Internet, blog posts I said I *totally* read but she knows I *totally* didn't, YouTube videos of corgi puppies...you know, the usual stuff modern couples send around thanks to the joys of technology and the Internet. And then, when I typed in this mysterious other email I had been using, I discovered the cause of all the confusion.

We had emailed each other several years ago.

And we'd only first met last year.

All right, let's break the heart of Heinlein and his *Door Into the Summer*, and let's take a door into the *winter* instead. Specifically, the late, fading winter. January of 2007, when I first moved from New Jersey to the city of Philadelphia for graduate school. I only knew two people in the city, one of my best friends from high school and a pal from my days of playing saxophone in ska (yes, yes I know) bands. The first would help me buy furniture for my first apartment, teach me how to cook, and introduce me to the joys of beer before moving back to her (and our) hometown, while the second... well, we are groomsmen in each other's future weddings, so needless to say we are still close.

However, I was still pretty lonely in a new city all on my own. So I did what a lot of people do...or at least did, back then.

I turned to Craigslist.

There, in the website equivalent of a dark alley, I posted an ad in the Strictly Platonic section hoping to connect with people like me, looking for friends in a city that felt big and intimidating. I hadn't heard of Meetup yet, and I think I used a T-Mobile Sidekick to post my ad. It was an early one, black and white screen and all. And that's how we first found each other.

The emails back and forth were innocent enough. We talked about our goals. She wanted to become a probation officer, while I was a poor graduate student dreaming of the novels I'd write someday. She spent her evenings dining in fancy restaurants and partying in nightclubs, while I was surviving on a diet of pasta and the dollar menu at Wendy's, going through the first two seasons of *Supernatural* on DVD. We swapped AOL Instant Messenger names (remember when that was still a thing?), and I'm not sure what happened after that.

We never met up.

Was it something one of us had said? Did I come on too strong, or did she simply lose interest in this odd stranger on the Internet? Maybe I left a weird away message up. This was 2007 after all, and that was still a thing people did. Perhaps it was due to the likely silly lyrics from a New Found Glory or Taking Back Sunday or Dashboard Confessional song, and she opted to stop talking with me.

Maybe I mentioned the whole *Supernatural* thing? The fact that I cut up dollar menu chicken nuggets and put them in the dollar menu Caesar salad, thus creating the cheapest and, frankly, most brilliantly conceived chicken Caesar salad in the downtown Philadelphia area?

Whatever the case was, it never happened. We never

got together. Had technology in 2007 failed me...or perhaps, done me a favor?

Let's push the lever forward on Wells's time machine and skip to the year 2012, summertime. I've left the world of Craigslist for a more updated visual platform: OkCupid. Here, people post pictures and lengthy biographies that detail just about everything. Hair color. Favorite music. Height. Career goals. Salary. You can even rank people with stars, as though they're books, computers, or toasters on Amazon.

It's here where we reconnect thanks to a supremely awkward message sent her way, followed by an array of notes sent back and forth. Sometimes from my computer, other times from an app on my iPhone.

We talk about our careers, our goals, many of which have changed if one were to compare the original Craigslist emails with our OkCupid messages. But most of the core stuff is still there: working with people, writing, graduate schools, and the like. We leave the constraints of OkCupid messages and quickly start sending emails, text messages, and quick, nervous phone calls, our letters and voices carrying through the air and space, helping us find one another again. Even though we didn't know we'd already found each other so many years ago.

Push that time machine's lever up four months. It's here, around my thirtieth birthday, that I realize this is the woman I'm going to marry. I first bring this up in my favorite bar, where I'm chatting with one of my best friends, the guy who used to play in ska bands with me, while sipping on a Yuengling, the beer my hometown pal introduced me to...as though everything had come full circle.

A little over a year later, she'd say yes.

When I tell this story, most people (myself included) make remarks about how it would have been amazing if we met back then. But back then, in 2007, we probably weren't right for each other. Over the course of those years, I learned a lot about myself. About the man I wanted to be. As for her, she moved a lot, finding herself in new cities on new adventures, learning more about who she was before coming back to settle in Philadelphia.

Who would she have met in 2007? That version of myself is someone I hardly recognize anymore. The struggling graduate student, still so torn about whether or not he'd actually make it in this new state, in this new city. Living in a two-bedroom apartment with three strangers, one of whom slept on a bare mattress in the living room. Still, 2007 me had a lush head of hair. There was that. But that's all superficial. The important stuff, like having mature relationships, discovering and achieving major personal goals, making an effort to connect with my family more, learning what it means to truly live on one's own...all those complicated things that make you an adult, were still nonexistent.

I think in this way, technology didn't fail me when it came to making a proper connection. It's a connection that I missed, but not a connection that I *missed*, if you catch my *Office-Space*-style phrasing. In some strange, unknown way, it pushed us apart, only to bring us back together at the exact right moment. The years gave us time to grow, and as I sit back and reflect on them, I find that I don't regret that time we had apart. In the end, it made me a better man for it, and a better man for her.

So maybe somewhere out there, somewhere in the future, time travel has been invented. Maybe when I'm old and fading, the technology will surface. And maybe I'll think back to this little essay, to our strange and quirky love story...and I actually won't write that letter, that Missed Connections post written for the me of the past.

The lonely version of me, in early 2007, probably would have received this note about my future wife and chased her with reckless abandon, ruining all chances of it working, leading to the non-existence of my future, happy self. He wouldn't complicate friendships, make mistakes, have regrets, and learn the lessons that shape him for the better. He wouldn't go on to have his heart broken in ways that paved the path for my future wife to mend it, to fix me, as it were.

Perhaps this essay isn't so much about wanting to send a note to the past, to post a blip on Craigslist, but to send this note to the future. A reminder to be grateful for your time apart, to remember you met at the right time, in the right place. That all the heartache and tears...it'll add up to something. And that something, that someone, will show up, when you're finally ready.

And on that note, we'll push the lever of this time machine to 2014, coming back to the present, where everything is, thankfully, as it should be.

ONE NIGHT ON ROT

Jonathan Farbowitz

Under cover of current,
Legions hold dances through the night, the black hat and the white
Nothing but signal, ineffable true static
Although at 09:00 GMT on Rot nothing is static.
Hijacking mail messages intended for, who knows,
Takedowns of information, *in situ* expurgation

Plunge to the underage underground
Ribald phrases exchanged for cyberbrothers
Taking shots of chatter
GIFable exchanges
Deprecated, out-of-business borders

Faceless in a flock
Slipping between three-fold loose worlds
Stability in a machine
More than that, a confidant.

The State deployed their useful bots, inserted and multiplying
Chose to throw it all down at 09:00 GMT
In darkness through North America.
The call went out: solidarity logons through the night
Rushing the ramparts, making sockets sore

So have the layers been lost
Or are data packets still slaloming the pylons?

The Music of Evolution

(or, A call-and-response with Infinity wherein a musician grapples with technology, gives up the ghost of tradition, and learns to love the laptop)

Sean Hoots

Ex Nihilo Nihil Fit

Nothing comes from nothing.

The *zero-energy universe theory* proposes that the total amount of energy present in the entirety of the universe equals exactly zero.

The number zero is not nothing.

Rather, zero merely represents an empty set. It can be used to add, subtract, divide, and multiply, in each case achieving a satisfyingly real number.

To uncover the full value of zero, the null set must be viewed as a point of universal balance, the precise fulcrum about which all of existence both fluctuates and stabilizes. On the side most immediately visible to humans stands matter. On the other side, the great invisible tugging force of gravity pulls the beam of being back to zero.

Heavy Metal Complexion

Art theory is not my bag. I am an artist, so I suppose that affords me some perspective or perhaps gives me an implicit kind of personal theory of art, but when it comes to defining it writ large, I'm more of a doer than a talker (or writer). My academic background may suggest otherwise, but I've been involved in the making of art a lot longer than I've wrestled with its philosophical pursuits.

Music has always been my medium of choice. Even when I ventured into other areas of expression, I couldn't shake my connection to music. Doodles of imagined (and unfortunately-for-them-never-realized) Slayer album covers? I have sketchbooks full of 'em. Half-assed plywood replica of a guitar upon which to shred maniacally while blasting Metallica in my bedroom? You bet I rocked it. Medium be damned, music always seemed to incorporate itself into the fabric of whatever creative endeavor I attempted.

(Full disclosure: I had a rather fierce heavy metal phase as a youngster in the '80s. Maybe it was all those pimples?)

Regardless of genre, I have always connected deeply with music. Every step from as far back as I can remember has been in the direction of that sound. I've dabbled in metal, blues, jazz, rock, folk, and electronic music, each to varying ends and effects, but always reinforcing that innate connection. No, that sound in my head has never been a sound in any specific style, but a broader sense of absolute involuntary frequency response to the world around me. An overwhelming feeling that, basically, just is. It's a gut-level punch, a tingling vibration that seems to radiate from my every atom, a flood of endless waves that come from nowhere, alternately crashing and soothing, and then ebbing back into the abyss...

Feelings are hard to describe. Words fail. That's why I play music.

The Lie That Reveals the Truth

Author Madeleine L'Engle proposes that art is communication, and for a writer that certainly follows, but the idea extends just as sturdily to other forms. Unpacking the process of art making, there seems to be a very human urge to process individual experiences in the world by expressing through some creative means a way to relate those experiences to others, be they real or imaginary. We see; we feel; we create (to explain what we see and feel). And then we cock our ears to the world and wait for some kind of reply.

These responses come in a variety of forms, from the ecstatic spiritual communion that occurs in the solitude of a moment shared with nothing but the vibration of a drone to the hysterical mass engagement that erupts in the presence of thousands of screaming rock fans. A dialogue exists between artist and audience, even if that audience is comprised of nothing more than the artist's own reflection.

As far back as the ancient Greeks, art was a matter of philosophical discourse. Plato and Aristotle spoke of it in terms of mimicry (mimesis), both to different ends. Plato saw art as an imitation of reality, and as such, it was a falsehood and therefore unworthy of pursuit or consideration. He had no time for distractions from his ideals.

Aristotle, on the other hand, found value in mimicry. He saw an opportunity for catharsis. By expressing themes common to an audience, the artist connects with members thereof and their personal take on the given theme, creating a conversation that may inspire insight and growth.

Yes, mimicry is essentially an imitation of something, but reframed, that imitation reveals itself to be a response on the part of the artist and, as such, serves a valuable function. The artist's response elicits a further response from the audience, and the dialogue has begun.

Picasso said, "Art is the lie that reveals the truth." It is a form that delivers a function. If nothing is created in a vacuum, then the conceit of art is not so much a falsehood as it is a different paradigm through which we might touch our feelings, and thereby the universe, in more profound ways.

If it feels right, do it.

An Uneasy Peace

Technology is perhaps the most obvious extension of human evolution into modern times. Anatomically, the species more or less settled its bones about 200,000 years ago (natural selection may yet persist in the physical dimension, but in subtler ways). Culture and language developed approximately 50,000 years ago, as well as technology in primitive forms.

With culture and language came art. Archaeologists have discovered cave paintings that date back 40,000 years. Using pigments derived from dirt, stone, and wood lying around them, early humans scratched images of animals onto rock walls that remain to this day.

From the same time period, flutes fashioned from vulture wing bones have been discovered, however, the music played thereupon is both academically and personally unknowable. We know music existed because we see tools fashioned to fit the purpose, but we have no songs. Alas, there is no Golden Oldies collection of the prehistoric era.

The oldest song fragments available come to us much in the same way as early history and myths have, via oral tradition. Eventually the development of language into written forms made its way around to music, and by 2,000 BC rudimentary musical notation appears on cuneiform tablets. The lines and notes we are used to seeing these days originated with European classical music only about 1,000 years ago.

Still, even upon achieving written tangibility, the art of music was subservient to the limitations of the page. Written notation gives a general roadmap of how a song should go, but oftentimes the composer of said piece is neither the conductor nor the performer in a particular reading of it, which can easily lead to misinterpretation. Surely, making someone else's piece of music "your own" can be rewarding in its own right, but that's beside the point at hand.

For example, Beethoven isn't immediately involved in modern productions of his symphonies, but they still resonate with enthralled audiences the world over. But what if we had the original version of his 9th symphony available for listening? Not the first public performance of it, but the piece as it sounded in the composer's head before it even manifested itself as ink on paper? Even beyond the fact that Beethoven had largely lost his hearing by his mid-twenties, might there have been certain personal nuances embedded within the work that could never be translated via musical notation? What if a note he intended within a given melodic strand wasn't quite an A but neither was it an A sharp, rather, something in between, indefinable by standard notation?

Eliminating all middling factors in the presentation of art gives the artist the best chance of getting the point across. As Susan Sontag notes in her essay "Against Interpretation," "...transparence means experiencing the luminousness of the thing in itself, of things being what they are." Increasing the transparency of the artist places all focus on the creation.

To our modern ear, most of the content of the history of music is unavailable without the filter of auxiliary means of reproduction.

Cooks in the Kitchen

While my primary artistic focus is music, I do find quite a bit of creative satisfaction in the preparation of food. Just like music, I have little to no instructional background in the culinary sciences. I just started messing around with foods that I liked one day, and the results have kept me coming back to the kitchen on a daily basis ever since. Improvisation, as in music, yields fruitful experience.

On occasion I've followed a recipe or two. If I'm aiming for something very specific, say, hollandaise, I'll do a little research to discover traditional methods of preparation, but inevitably, once I get to the actual work of cooking, all bets are off. I begin a kind of dialogue with the ingredients, letting them talk among themselves, listening patiently and then turning the whole conversation into a collaborative piece. The results are almost always tastier than any recipe I've ever followed, even if the kitchen does end up a bit messier for the experimentation.

A recipe serves as a kind of theoretical notation for the idea of a particular food. The actual process of preparing food, however, is what actually brings it into being—makes it real, gives it substance (and, if you're doing it right, a harmony of flavor). There is a tangibility and, ultimately, satisfaction inherent in the food itself. To my knowledge, no belly has ever been filled by reading a cookbook.

I find the recording studio to be very much like the kitchen—instruments lying about like so many veggies and herbs, meaty melodies slow roasting to perfection, and a big old blender in the way of a mixing console. A song can quickly find its fullness in the studio. What's more, the modern studio provides a forgiving environment in which the ambitious amateur can create through a process of trial and error until perfection is attained.

From Thomas Edison's earliest attempts to contain music in wax cylinders to the latest version of Pro Tools software, every iteration of recording technology brings a greater number of composers closer to crafting a clearer picture in the studio of what they hear in their own heads. Current computer software can replicate any number of instruments, both traditional and otherworldly, effectively eliminating the obstacles of physical availability (not to mention exorbitant cost and storage) and formal training, opening wide the gates to aspiring artists. The field is being leveled.

In his famous 1979 lecture "The Studio as Compositional Tool," Brian Eno admits that studio technology has afforded him, a musician who can neither read nor write notational music (nor play any instrument with proficiency, for that matter), the opportunity to compose. And if you're at all familiar with his work then you know how elaborate, elegant, and essentially game-changing his repertoire is. Prior to the advent of such technology, a guy like Eno would likely never contribute to the world of music in such a seminal way.

Just as my rogue kitchen experiments have yielded meals on par with "professional" restaurants, so too can the modern home studio make a legitimate recording artist of those with little more than vision and drive. Improvisation, trial-and-error, forging one's own path—these are the methods for finding the voice that will most clearly express what's itching to get out. Academic theory and privileged access to limited resources, such as costly, world-class vintage instruments and the virtuosos to play them, no longer prevent the amateur everyman from chasing down that sound banging around in his head.

By the way, you really should try my hollandaise. Just ask my wife. It sealed the deal with my breakfast-in-bed marriage proposal.

Swimming in the Sound

(The following is a poem-log tracing the creation of a piece of music using software on a laptop.)

Friday, January 10, 10:55 a.m. MST; Park City, UT

Home is Philly
But here is now
Rocky mountains
4 days of
Slipping down slopes
Muscles ache
In body and family
Under tension and pressure
Of relief and strain

Vacation is a different kind of work

Final breakfast; final day
Pull away to fill ears with something
Finding tones and textures
To fully process the grandeur
The gluttony
The glib agreements

It starts with the low end

Today needs an anchor
Synth module selected
Map a simple line
4/4
32 beats
8 bars of 4
Key of D
Good as any

LISTEN TO
SWIMMING
IN SOUND

[soundcloud.com/sean-hoots/swimming-in-sound]

Add an E and a G for interest
Chop off the last 4 beats,
Now a 28 step line
Tweak tone with onboard parameters
Virtual knobs and buttons
Point and click
Sliding sound
From ominous to otherworldly
A phase shift occurs in my head
Hear the line as starting halfway through
First half now the second half
Shifting line, shifting key
D becomes G
Slow tempo from 120 bpm to 80 bpm
Distress the signal,
What in life is perfect?
What is whole?
What is clean?

A bit of percussion nails the foundation to the ground
Drum machine module,
A snare in my head

Cracking on the 5 of 16,
Looped over 28
Shifting rhythmic tectonics

Pads and lines intertwine,
Elementary knowledge of the G major
Swoops, swirls and
Original bass line
(now interfering with more interesting elements)
Needs to transform
Raise the octave,
Change the sound,
Now a drone
Once an anchor
Now the sea

Building from start to finish,
A story
Fade in
Chill, squiggles and a punctuating pulse
Like something is coming from nothing

First lead voice
(known only as "distorted disco" to the module...
What is this instrument?
Who knows?
A kind of fuzzy, percussive sound
Plucking the fruit of creation)

Second lead voice,
A piano,

Heavily treated by various effects
When is a piano not a piano?

Kicks and a paper thin snare loop
Carving out rhythm,
Granting context
To drones, manipulations and swishes
This is where I am living
Here I swim in sound

Time to leave
Save progress

4:41 p.m. MST; Delta Flight 1114, Row 8, Seat C

Boarded and bound for home
80 bpm crawls to 60 bpm
Pacing reflects desire
A few nudges
A few clicks
Morphing arrangement
Finding its right place
Put the whole thing on loop
And settle into headphone headspace
Everything is slowing
To solidity

Sunday, January 12, 5:38 p.m. EST; Philadelphia, PA

Home, settled
Unpacking
Revealing track left waiting
24 hours of compression

A string section begs to take shape
Choir voices and flutes
New subtleties add dimension and shading
A thousand voices corralled in mere moments
An orchestra constructed of thin air

Infinity Mirrors

In 1984 Kurzweil Music Systems released their first synthesizer, the K250, to international acclaim. It replicated grand piano and various orchestral sounds so faithfully, most listeners could not differentiate the synth sounds from their real-life acoustic counterparts. The man behind this giant leap forward in electronic music, Raymond Kurzweil, is also a well-regarded author, futurist, and current director of engineering at Google.

Wait...futurist?

Indeed. Futurists, or futurologists, are typically scientists who attempt to explore the path ahead for the planet, considering all manner of variables, from economics to technology to weather patterns. Part of what Kurzweil and other futurists predict for human development falls under another oddly termed umbrella: transhumanism.

Transhumanism ideology suggests a kind of co-evolution of humans and technology toward a convergence at some point in the future, a "singularity," after which human and machine will operate as one. The exponential acceleration of technology over the course of human history suggests such a trajectory, and in fact, with each technological innovation, we continue to push beyond prior comprehension of what once were the limits of human capability and potential.

Whether or not you agree 100% with this forecast, you'd

be hard pressed to deny that our lives have become increasingly dependent upon and centered around technology. A vibrating invisible web envelops the globe at present, putting the universe at our fingertips. Nothing less than the sum total of all human knowledge can be accessed by the telephone that fits in your pocket. We are simultaneously gathering ever-vaster troves of information and storing them in ever-shrinking trunks.

Unlike the typical human process of knowledge acquisition (i.e. learning, training, apprenticing, practicing), computers gain skills and insight by simply downloading coded patterns of 1s and 0s in a matter of seconds. Once these binary codes are acquired, they can be shared instantaneously with billions of other machines. If technology is indeed moving toward a singularity with humankind, then it isn't hard to foresee a time when humans will consume and utilize information in a more immediate way as a matter of evolutionary course, eventually making all things known to all people at all times.

Wired magazine co-founder Kevin Kelly presents the scenario with an elegantly evolutionary perspective:

> *As evolution evolves, it keeps piling on different ways to adapt and learn until eventually the minds of animals are caught in self-awareness. This self-awareness thinks up more minds, and together a universe of minds transcends all previous limits. The destiny of this collective mind is to expand imagination in all directions until it is no longer solitary but reflects the infinite.*

Everything is Everything

Back to the top, nothing comes from nothing. The energy that fuels everything around us exists in infinite supply as the universe self-balances between the substance of matter and the force of gravity. The drive to create, whether technologically or artistically, defines who we are as well as what we are becoming. In the reciprocal relationship between the fleshy substance of our being and the transformative force of our ambitions, creator and creation co-evolve.

The trajectory of technological innovation, especially over the last century, seems to be guiding a concurrent, interactive evolution between man and machine. Music technology in particular frees the would-be composer from old-world constraints, reducing a millennium of educational constrictions and financial implications to a handful of mere clicks on a single computer. We are collapsing time and space, as the artist is freed to explore more of the universe.

If we are indeed headed for a singularity in which we all share all knowledge all of the time, how much more comprehensive will the dialogue between artist, art, and audience become? How much more transparently will we be able to share our creations? And how much faster will the responses come?

And when these responses come so completely and immediately as to, in effect, serve as reflections of one another, as per Mr. Kelly, what becomes of personal identity? Will dissolution of the individual lead to the emergence of a mass of pure, sentient creation, a call-and-response feedback loop? When artists become truly unfettered and released into the cosmos, how far will we roam without the need of stultifying body and mind?

Like So Many Stars

Alex Vuocolo

When the famous psychologist and philosopher Dr. Anna Refn died, there were explicit instructions in her will to tell the public that she disappeared while hiking the windy mountain trails of Nepal.

The truth is she died while watching television after a long evening of flicking through the channels, scanning for faces in the static.

It should be noted that the last time she visited Nepal, ten years prior, she returned with a mental technique for better empathizing with other human beings through technological mediums—a task that Anna believed was increasingly important for the peace and continuity of the human race.

"In such an interconnected world, mobilizing our consciences is as important as sharing our resources and capital," Anna wrote in an essay for a social sciences journal. "It's day-to-day work now, empathizing with our far-flung cousins. We can't depend on the catalysts of tragedy to remind us of each other's existence."

"So what are the alternatives?" One of her intellectual rivals asked. "Should we buy plane tickets for all the world and hold meet and greets?"

"No," said Anna. She insisted that the overwhelming majority of people would only see their distant human cousins through a television or computer screen. She knew this because her own love for other people and cultures began like so many of her generation: in front of a back-lit screen.

"What are the most common cultural transactions in

the world right now?" Anna asked an audience of freshman anthropology students. "The kind that occur between a screen and a viewer, and it is at that juncture that we must intervene."

In effect, Anna's technique aimed to leverage these small, continuous interactions by building upon the empathy that was already there, almost like adding rebar to concrete.

Anna, who was always keen on using mechanical or scientific metaphors to describe psychology, called her technique a new form of "thought-machinery," which supplemented the mind in the way a magnifying glass supplements an eye.

"It's not such a strange concept," she would say. "Consider the fable, the riddle, the math equation, all of these are thought-machines. They are constructs in our heads that execute tasks, that order the chaos of our minds and have as much potential for development as any circuit board or industrial tool."

Following that logic, Anna refined her technique in the solitude of Nepal, tinkering like a gadgeteer until it was as polished and accessible as any mass market toy.

She returned from her reclusion with a comprehensive strategy for training her technique to the rest of the world. It involved mass-produced hard copies of video lectures, syndicated radio shows, television appearances, speaking tours, op-eds.

By the time Anna finally had a chance to look back and observe what she had brought on the world, it had already happened: distant cultures were cooperating. They were growing kinder and learning to see the dependencies among one other. They were using technologies to understand rather than mythologize or propagandize.

This effect did not surprise Anna. She was confident in what she made. What she didn't anticipate was the monumental love that flowered, not just generally, but for herself. She had become almost like a saint, except thoroughly modern and secular.

She was loved through screens and countless other mediums, some as minimal as clipped sound bites of her voice, heard like echoes across mountain valleys.

As she aged, this affection only intensified, and she even began to fear that her death would unsettle the millions of people who had taken solace in her teachings.

These fears now seem overstated. But what about the secret fear that Anna held quietly in her heart? What about the possibility that her thought-machines had cleared the barriers of shared experience to such an extent that sadness and loss coursed through them like a cold wind?

...

The first announcement of Anna's death came over an international news site, then trickled down through the ducts of the global information network, at last sputtering through the speakers of hand-cranked FM radios in remote villages and mining camps.

In a rich suburb of Mumbai, a teenage girl walked down a manicured path towards her house. She heard the news through earbuds, which fed into her mind the day's headlines and weather reports in a continuous stream. When she heard the clear and yet devastating headline, "Dr. Anna Refn Dead at 67," she stopped walking. Then came the recording of Anna's voice, and the girl nearly crumbled.

Simultaneously, an international chat room became abuzz with badly-spelled pronouncements of sadness and admiration. "*You changed my life, Anna*," said so many messages in so many words. Audio files, slivers of video, and images of Anna coursed between the mourner's networks, forming a kind of branching, digital funeral procession.

In a cafe in London, a group of students took each other's hands and did their best to lend sacredness to that moment by gathering around each other to watch one of Anna's most famous lectures. Anna stood at small podium in Central Park. The sky was bright behind her and the wind swept through her short brown hair. The camera was close, and the students took great pains to follow the movement of her lips as she spoke.

The patrons of a Philadelphia dive bar gathered around the radio as they drained pints of beer and listened to an old and very famous interview with Anna that circulated in the day's following the announcement of her death:

"You realize, Dr. Refn, that you have become the greatest beneficiary of your technique?" the interviewer asked Anna.

It disconcerted her to hear this, but after a moment of thought she found reason in the suggestion to at least grin.

Anna responded: "Our mothers and fathers are the first to love us, and for a long time they are the only people we know how to truly love. So perhaps as I taught my method, some people found their passion for the practice in their passion for the teacher."

And then, of course, the anchor asked: "So what happens when the teacher is gone?"

She smiled and then gracefully changed the topic. But in truth the question haunted her as she left New York City that evening, bound for her solitary home on Maine's southern coast.

It was around that time that she became scarce in the public's view and began writing a novel, which she hoped would imbue her ideas into something longer-lasting than her own persona. She also began planning her death. She hoped that the fake trip to Nepal might shroud her death in mystery and diffuse the stone-hard finality of finding an old woman's body sunken into a reclining chair.

But her plans were spoiled. Her body was found by a young neighbor who often visited for coffee. In line with the rest of her generation, this young woman was compelled to channel her shock and fear into a public forum.

Hardly a day later, the rest of the world knew that Anna was irrefutably dead. There was no cushion of uncertainty, no mystical Nepalese landscape where Anna's legacy might freeze over and persist through the ages as a changeless totem.

...

Astronomers and stargazers alike have attested to the power of observing the cosmos—how it overwhelms, affirms, and terrifies all at once. It is easy to suggest that the source of these feelings is the sheer size of the universe. But for Anna, to observe another human being had a similar effect. "It's like seeing so many stars," she said.

She reiterated this idea many times, using the metaphor of the cosmos to convey the vastness of a single person. She explained in her lectures that the problem was the seeming lack of spectacle in the human face. While the stars and nebulae put on these ceaseless, dazzling performances, the human face expresses itself in subtle movements and coloration. Sometimes a face can prove to be monstrously expressive, but then the problem reverses; it becomes like staring at the sun, too intense to fully manage.

Through technological mediums, this effect is further complicated. A human face can be close and gigantic, a field of pores. But it can also appear so small and muddled that it seems to be no more than smudged pixels.

It is difficult to summarize how Anna's technique addressed this problem. Like certain clever knots, the emergent effect comes after a confusing sequence of directions. Loop the end of the rope through the hole. Pinch the line as you pull it tight. Then there it is, an improved empathic sense.

In the crudest sense, the technique generates empathy from any beginning point, however diluted or small. If a face is abstracted by distance, blurred and mutilated by the resolution of a screen, then a certain amount of rebuilding is required to find the full humanity behind that face—a piecemeal construction that begins with that single broken sample.

When it works, it's astounding. A feeling washes over you and suddenly the immense gaps in space between yourself and another person feel smaller. They became crowded with the memories and feelings of the other person as if they are sprawling out into space with each new thought.

Anna had spread her love for people far and wide, through telephone wires, radio signals, bands of fiber-optics, and the people who shared her technique by word of mouth.

But what happened to Anna's great infrastructure of empathy after she died? Was the news anchor correct to suggest that this global bout of empathy was hinging on the narrow shoulders of an old philosopher?

Near the end of her life, Anna considered these questions extensively. She worried and worried, deep into the night sometimes. She listened to the waves crash against the rocks outside her home and dreamed of the world, spinning its entire shape in her mind as if trying to encapsulate it in a single thought.

She feared so much that her chest ached, and she knew only vaguely that her fear was the pinnacle of the kind of empathy she tried to cultivate with her technique, a vast love that twinges with every death and hardship across the globe.

At last, she resigned herself to it and watched the faces of the people on screen, exercising her ability to love, finding kinks and new strengths in her thought-machines, building upon them even as she began to dream.

ON **OTHER BELTS, OTHER WORLDS**

Gravity and the Cloud

Miriam Seidel

As I watched Sandra Bullock dodge flying satellite debris in the film *Gravity*, the slowly unfolding background panorama overtook the space-thriller story and became the main event for me: brilliant digital renderings of cloud systems, coastlines, and especially the thin, blue-white film of the atmosphere coating the Earth's surface. There they were: the biosphere and the atmosphere, as seen from space for real by just a few hundred people so far.

That delicate, translucent outer membrane of the atmosphere offers a barely visible analogue to other, unseen spheres. Is this what cyberspace would look like if we could see it? After all, we're now calling it the Cloud, an atmospheric image that seems to have won out over earlier metaphors, including the highway, the ocean, and even the web. We know that cyberspace constitutes a reality separate from its physical plumbing—the computers, servers, cables, and satellites that make it possible.

Then there's the nöosphere, a concept developed in the 1920s—long before the digital revolution—by the French theologian Pierre Teilhard de Chardin, out of conversations with Vladimir Verdansky, the Russian scientist who himself coined the term "biosphere," and Edouard le Roy, a French philosopher. Like cyberspace, the nöosphere is invisible. It is a kind of thought experiment, a part of Teilhard's global vision of human evolution: imagine that all of human thought surrounds the Earth in an invisible shell. As our mental outpourings grow and intensify, this "thinking layer" (the prefix comes from the Greek *nous*, for mind) fills in and comes into its own.

Teilhard suggested that the nöosphere would emerge out of technologies "extending a closely interdependent network" around the world. He was referring to radio, teletype, and television—a web of communication that seems to eerily anticipate the Internet and our current web of digital communication. A Jesuit priest, he saw the nöosphere in Christian terms, as moving toward a fulfillment of divine consciousness in the world. He was prevented from publishing his theories by the Catholic hierarchy—yes, this happened in the twentieth century—and none of his books appeared until after his death in 1955. Since then, the nöosphere has provided a tantalizing paradigm far beyond the parameters of Catholic theology.

Another invisible mass extending beyond Earth's surface, the magnetosphere, has also been suggested as a nöosphere-like entity. It includes the Van Allen Belt, a doughnut-shaped whorl of charged particles anchored by the north and south magnetic poles. The full magnetosphere extends several Earth-diameters out around us. Before James Van Allen identified the Van Allen Belt in 1958, the American physicist Oliver Reiser speculated on its existence. Reiser proposed that huge radiation bodies surrounding Earth might act as what he called a PSI bank: a repository for human thoughts. This offers an intriguing new shape for Teilhard de Chardin's thought experiment.

We're still learning about the magnetosphere, with new information coming in from satellite probes just this past year. Recent findings show what seem to be great magnetic windstorms within the belts: "large-scale, global motion involving ultra low-frequency waves pulsing through Earth's magnetosphere and operating across vast distances up to hundreds of thousands of kilometers" (Harlan Spence, University of New Hampshire). Massive electromagnetic weather fronts circling above Earth? This brings the concept of "the cloud" into a whole new realm. As a metaphor for the collective consciousness of the cybersphere, this image is far wilder and more mysterious than, say, the early inside-the-silicon-chip world of *TRON*. It makes me wish we could honor the Van Allen Belt with a name that captures its deep strangeness a little better, just as Chomolungma (Mother Goddess of the World) sounds so much more fitting than Mount Everest.

A space dork (and I count myself in) will find all these spheres and whirling belts intrinsically interesting. But they're also worth contemplating as a way to help understand the truly bizarre new development of our digital environment. How strange is it that we, a large piece of the world's population, have eagerly, voluntarily immersed ourselves in a virtual realm that didn't exist a half century ago? What is it we're experiencing? "Virtual" and "cyber" and "cloud-based" are shorthand terms allowing us to label something that we haven't begun to grasp, even as we surround ourselves in it. The speed of our adaptation to this breaking phenomenon has been so fast, and its scale so great, it's almost as if we have been hypnotized by our irresistible new powers. We have failed to pause and wonder at it, except maybe in the realm of science fiction.

Yves Klein, the French conceptual artist, envisioned a new "Architecture of the Air" in a series of drawings made not long before he died in 1962. In the future, he proposed, human beings would be liberated from opaque walls and even clothing. He imagined people living communally in huge shelters, protected from cold and rain with invisible roofs made of jets of compressed air. Klein, who called himself "Impresario of the Void," had taken the sky as his muse long before this, adding his non-physical artist's "signature" to the sky and creating his own pigment, the sky-hued International Yves Klein Blue, as an iconic color representing the Void. In some early art happenings, he made action paintings by pressing nude models coated in Yves Klein Blue to the canvas, yielding haunting images of life-sized, floating sky-colored figures released from gravity. Klein's utopian vision may be someone else's nightmare: the untethered astronauts of *Gravity* are justifiably afraid of floating out into the void of space.

Still, with his flying bodies and his imagined walls of air, was Klein tuning in to a poetic vision of the future that somehow rhymes with the phenomenology of our current digitized world? We're living a big part of our lives inside the Cloud now, and experiencing vertiginous change there. Among other major shifts, our sense of privacy is radically morphing as we experience this new reality, with its transparent structure of digital wind. In the movie, Sandra Bullock manages to make it back to solid ground after a horrifying near-death in unending airless space. We're not about to float off the reassuring soil under our feet, but our new sphere of experience is taking place in an odd, electromagnetically-charged hybrid zone, neither physical nor nonphysical, earth nor heaven. We're all astronauts exploring a new plane, one that we're just beginning to understand.

Immunity

Christine Neulieb

The control panel blinked serenely under Om's ancient hands. The skin was smooth and fresh like that of a twenty-year-old child in the olden days, before AntiAge had come. But he had lived seventy years aboard the ship, and twenty-eight before that on the First Planet.

Every switch he flicked made an echo in the solitude of the chamber. By now he could steer the ship almost in his sleep. He had memorized not only the rocket science that he knew when he came on board as a junior engineer, but also every line of every science book in the ship's library. In lonely hours, he solved quantum equations that had puzzled the men who *built* the starship. Perhaps they still puzzled over them; there was no way of knowing. The ship had passed beyond the range of Earth communication decades ago, by ship-time. More than a century ago, by Earth-time, but Om preferred to think of himself as a healthy, young ninety-eight years old, not 172. By *spacetime*, he swore under his breath, he was not 172.

The massive spooky-propulsion rockets that drove the ship were bright on the screen beneath him, though radiation shields hid them from the naked eye. They stretched out into the cavernous blackness of space like a monster squid and its tentacles, filaments pulsing with light. All seemed in order.

When he had completed the final checks of every system, he clambered out of the control room and down the dim hallway to his quarters, making a detour through the library as he did every day. There was always something new to read, and when one had decades to read, that was a mercy.

Just yesterday he had scrolled through ephemera from the twenty-first century and found a copy of the very first advertisement for a commercial space flight. It spoke of a cataclysmic—yet perfectly controlled—burst of power, surging through the ship when the rockets fired. It tempted passengers to the terrifying thrill of sudden acceleration and screaming motors. The sky beyond the windows would swiftly change color, brilliant blue fading to indigo-violet and then, at last, to that cavernous black. You are weightless; you catch a priceless glimpse of Earth, the ancestral home of humanity, from space.

It was doggerel poetry meant to seduce the super-rich.

By the time the Hesperides Corporation invented spooky propulsion, such trips no longer cost the equivalent of several years' salary to the average earthling. If not everyone had seen the sky darken to coal-black around them, at least everyone knew someone who had. They did in the still-powerful empires of the Pacific Rim, anyway. Om hailed from California.

He switched on the book templates and pulled up a treatise on the holographic principle, but the tingling in his legs and fingers made it hard to concentrate. The voyage was nearing its end at last; his body would not let him forget it. Tartessos, the planet they were steering for, had been a real, audible blip on the ship's sensors that day. It was the first sight of land to a sailor who'd been at sea for a lifetime.

On Earth, everyone had talked that way about man's first journey beyond the stars: *voyage, crew, ship, sailing, uncharted territory*. When Hesperides recruited colonists, they'd set it up so the process felt like signing onto the crew of a nineteenth-century whaling ship, headed into the vast unsafety of the Arctic seas. The walls of the library bore images not of the first space travelers but of ancient mariners. There were the seven-masted junk ships of the Han dynasty, the dragon longships of the Vikings with majestic curving prows, and the colorful fleet of Ferdinand Magellan. Another wall showed even more primitive fleets, those of the first human ancestors to navigate the Pacific islands. The tribesmen sailed terrified through a storm, until finally at the far end of the wall the waters calmed and they fished for their supper. Every image in the room was of water.

There was water on Tartessos. Water, breathable air, perhaps life. At the thought of fresh air, Om inhaled deeply, but of course all he tasted was the ship's canned air. Its plasticky odor had stopped bothering him a long time ago, though when he first got on board it had made him feel constantly slimy, like he wanted to scrub all his skin off. Suddenly he couldn't stand the ship's air any longer and couldn't understand how he'd been able to put up with it for so long. It would be a few days before the ship began to prepare itself for landing. Om gave up on reading and returned to the control deck, where he could listen to the precious beep on the ship's sensors of the nearing planet. The lights on board began to dim at 20:00, to simulate the setting of a sun. Om listened to the beep through the ship's night. Toward morning, he thought that he could hear it getting louder.

...

In essence the starship landed itself. Its coordinates had been programmed back on Earth, and it had taken constant measurements in the interim, as it was supposed to do. As the ship entered the Epsilon Indi system, her sensors noted when tiny asteroids dinged the hull. It was like a chorus of triangle players. Not once did the alarm signal danger—not one monster asteroid threatened to tear the ship apart. Tartessos approached. Om watched it, awed. Swirling clouds cloaked the planet's surface, but they looked gray-brown from space, which struck him with intense nostalgia for the pure white clouds of Earth. The ground beneath showed the unmistakable blue of water, but also streaks of angry purple. Om buckled himself into his seat ahead of the shock that would come when the ship jettisoned its energy reserves. After that it was his job to land her. He hoped he'd studied well.

After the ship had blazed through the atmosphere, he lifted the radiation shields for his first naked-eye look at the colony's new home—and saw *trees*. Not water clouded with primitive space algae, but trees, or something like them. They were much bigger than Earth trees, with succulent purple-red foliage drooping from the branches. The sight was such a shock that it broke Om's concentration for a perilous few seconds before he corrected his course and made for a high plateau with a wide, flat, rocky area perfect for landing.

Om paced back and forth in the laboratory while he waited for the results of a myriad of tests on the planet's air quality, surface temperature, chemical composition, and gravity. The ship was designed to function as a self-contained city if it wasn't possible to survive the outside environment without a spacesuit. Hesperides had cautioned everyone that it might not be possible to step outside for years. The corporation was afraid people would be so cabin sick as to rush outside without thinking, like a herd of lemmings.

Over and over, against all odds, the readings were Earth-like. Om couldn't believe it. There was enough oxygen to breathe, just barely. The atmospheric pressure was elevated, but not oppressive. It was hot outside, but not deadly. The heat of day would be like a vacation in the tropics. It wouldn't broil the meat off a man.

Om's footsteps echoed in the empty loading dock a few days later, as he prepared to open the hatch. Everything behind him was sealed off. With a trembling finger he pushed a button and the doors parted. A hot wind rushed in. It tasted of metal and decaying leaves. Om trudged to the mouth of the dock, feeling fifteen kilos heavier than he was used to. The fresh air stunned him so profoundly that he stood in the hatch without moving for almost an hour, before he remembered to set a boot on the cinnabar rocks outside.

"Another small step," he murmured, blinking up at the hazy orange sky. Then he went back inside to begin the task he feared the most.

...

At the heart of the ship was a huge mausoleum of cryogenic columns, thousands of them stretching in rows from port to starboard. Those columns held Om's shipmates, frozen in time some twenty years into the voyage, when they had started to forget things.

Hesperides Corp. had employed a battalion of the world's greatest scientific minds and set them to work in a rocket science playground. They had run every calculation hundreds of times, including how much shielding they would need to protect fragile human brains from decaying in the radiant energy of the cosmos. The living spaces onboard ship had walls of lead six feet thick, while magnetic rings scattered particles away from her hull. Moreover, AntiAge had sealed the colonists' brains against dementia and their bodies against cancers. Or so they had been told. Everyone's worry was that the planet would not, in the end, be as Earthlike as they hoped—that generations of colonists would have to live inside the sheltering hull of their starship before there was enough oxygen outside to breathe. Or that the temperature would be just slightly outside the habitable range. No one had anticipated that one year the colonists would simply begin forgetting each other's names, the location of the showers, what they'd had for breakfast.

The dementia plague accelerated so quickly that they began cryonic preservation of the worst cases six months after the first symptoms appeared. First it affected one wing of the ship more than the others, but then it began spreading everywhere. They were becoming a ship of fools. Even the medical and scientific teams lost their expertise. Instruction manuals that they hadn't needed to consult since they were students appeared all around the lab, dog-eared and underlined. Once Om walked through the lab and saw an open manual in which every single word had been underlined and highlighted. Later on he had nightmares about it.

Om seemed to be the only one unaffected. In desperation the scientists trained him to carry out the cryonic process, not realizing that they were explaining the same steps a dozen times over. He ordered them to start freezing people as soon as they had memory lapses, hoping to arrest the process of decay. Eventually he took over, freezing the scientific team and then the last few colonists on his own. Yet his own mind remained clear. For reasons unknown, he was immune.

"I don't know if I fit in this contraption," said the ship's first mate, the last one to be frozen, as Om shepherded him into his column. It was the last time Om had heard a human voice.

Ten years later, just to pass the time, he had sequenced his own DNA and found the rare genetic mutation that protected his squishy human brain from cosmic radiation. Imagine, he thought, the headlines in scientific journals back on Earth. He wrote up his findings and preserved them in the ship's library.

...

Now there was nothing to do but unfreeze the colonists and see what state they were in. Om had always held out a faint hope that the process of bathing the brain in nano-chemicals and then restoring it might reverse the effects of whatever had happened before. His shipmates would greet him with embraces, unable to believe the journey was over so soon. He tried to think only of that as he initiated the rapid-warming process. He could only do a few columns at a time, since cryonics had never been able to come up with a way of revivifying people that didn't involve a good old-fashioned shock to the heart.

The ship's first mate sucked in air and his eyes flew open, rolling wildly.

"Wha— wha— wha—" he cried, and stumbled to his knees. He blinked up at Om, showing no recognition.

Naturally it would be disorienting to open one's eyes again after being dead on a spaceship for fifty years, Om thought. He pulled his lab coat away from the man's fingers and moved on to the next few columns, where he revived a lab technician and a cook. They reacted the same way as the first mate: astonished as newborn children.

"It's okay," Om said weakly. "We're here. We've reached Tartessos."

"Tar. Sos," the lab technician coughed.

About 700 of the frozen colonists revived, all told. The rest showed no signs of life, not even when Om shocked them repeatedly. He shut the columns that held the ones who didn't revive and set the dials cool enough to preserve the bodies. The rest, after they stumbled around like zombies for a few hours, came to their senses enough to ransack the kitchens for food. They seemed to have the minds of three-year-old children. They knew they were space travelers; they did not remember being frozen; they did not remember the horror of knowing there was a plague on board. Some were worse off than others and couldn't seem to remember how their limbs worked. They shuffled up and down the corridors with mincing steps like wooden marionettes. Om went about persuading these ones to sit in wheelchairs, which seemed to calm them.

...

Om decided it would be kindest to let his imbecilic shipmates live out the rest of their natural lives on the ship, because outdoors they would wander off and be lost, helpless, unable to find food or shelter. He left the ship each day alone and began assembling the portable dwellings that had been packed in boxes in the cargo hold. He did not ask himself why he did this; perhaps he had not yet given up hope of training the demented ones to live in the open air. As he worked he recited to himself an ancient poem of Earth:

I will arise and go now, and go to Innisfree
And a small cabin build there, of clay and wattles made:
Nine bean-rows will I have there, a hive for the honeybees;
And live alone in the bee-loud glade.

...

Indoors it was pandemonium. Few of the colonists could be bothered to wear clothes or use toilets or eat with utensils. Cleaner robots ran constantly, which took care of most of the mess, but a faint hospital odor began to pervade the whole ship. At night the colonists slept in heaps on the sofas in the lounge, like tired dogs. Om checked the monitors for strays every evening and herded them back to where the rest of the people lived. Sometimes he found one as far away as the agricultural labs, but mostly they stuck together.

There were a few people, however, who gravitated to the windows, where they gazed out at the purple trees with longing. Again and again Om rounded them up and brought them back to the common rooms, but their faces became so wistful that he feared he was breaking their hearts by keeping them indoors. It wasn't a lot of people, just a couple of dozen, always the same ones, some of them in wheelchairs. No matter how many times he moved them, they returned to the windows. One day he found a girl—he thought that she was a mechanic, and that her name was Lydia—gazing motionless out a porthole in the corridor, so enthralled that he couldn't stand it.

He decided to experiment. He decided to round up the group of window-gazers and move them outside, into the little circle of huts that he had constructed with a fire pit in the middle to ward off the darkness of night, though there was no need for it on a planet that had no native animal life. He would train them to live in the houses and shell beans and not wander off. He would see if they *could* be trained.

It took hours to collect them all and walk or wheel them down to the loading dock. There they milled around with vacant expressions until he pushed the button to open the hatch, and then the air rushed in and their eyes sprang wide. The ones who could walk shuffled out and gaped at the new landscape. A few did strange gleeful dances or clapped their hands. The ones sitting down worked their mouths as if trying to eat the air. Om pushed the wheelchairs out onto the rocks, bumping over the uneven terrain to a place that was level enough to park them. There he lined them all up in a row and stood behind with folded arms to watch over his charges.

Lydia danced the most gleefully of all, capering across the dark-red rocks. Despite the long cryonic sleep she was surefooted. She wore a light summer dress made from cotton that had grown in the soil of Earth; the strong winds of Tartessos whipped it behind her as she danced.

After a few minutes, though, she seemed to become aware that she had ranged far beyond the others. She stopped dancing and returned to Om, looking up at him with a puzzled expression.

"Now?" she asked. "Here? We."

There was an absent, monotonous quality to her voice, a quality he knew well—it had come to afflict all the colonists, and it meant she would not comprehend an answer. She would never comprehend an answer. None of them would. The neurological damage was irreversible.

Om sighed, feeling very tired and very old. AntiAge

would not work on the new planet, where they were supposed to begin the process of natural reproduction. He looked around at the women, whose bodies were in the flush of youth but who were as mindless as ancient grandmothers in the olden days. One of them, sitting in a wheelchair, had begun to drool. They would, he supposed, be able to bear completely normal and healthy children. The thought of it turned his stomach. The primeval forests of Tartessos unfurled in the distance under a coral and tangerine sky; an early moon rose to hang above the majestic trees like a ripe pear. Om, alone, sat down on the alien ground and wept.

Life Elsewhere

Phillip W. Snyder

I, for one, want to know if the life we find in every nook and cranny of this planet constitutes something completely unique in the universe—as we seemingly take for granted—or whether life exists similarly in all the nooks and crannies of every planet in the universe where the conditions are right for the origin of life as we know it. That is a complicated wish, and as I continuously age, the likelihood of that wish being granted (before I'm gone) continuously diminishes. What I find hardest to accept is that granting that wish just might require a gargantuan effort of humanity: an effort much like putting a man on the moon and returning him safely to Earth, squared.

On the one hopeful hand, the evidence exists to make a case for asking, and humanity possesses almost all of the technology we would need to answer the question "Are we alone in the universe?" On the pragmatic other hand, humanity continues to grapple with global challenges of energy and water consumption that rightfully marginalize most conversation about making the effort to get answers. The pace at which we are exploring is painfully slow, even though understanding how life elsewhere has evolved to solve analogous challenges might lead to solutions at home. Although publicly funded space exploration instills little hope of success in this endeavor, hope remains that private collaborations may soon see the potential value of finding life elsewhere.

The current prediction—based on the observation of more than 1,000 confirmed and potential planet-harboring stars—is that more than 20% of the sun-like stars in the Milky Way harbor Earth-like planets. (I will come back to the idea of "Earth-like" in few paragraphs.) If that fraction holds, however, it means that billions of habitable planets populate the Milky Way, which is only one of perhaps hundreds of billions of galaxies in the visible universe. It thus seems quite possible that more places exist in the universe where life—like that we find on Earth—has evolved than there are atoms in a drop of water. Even if the spark of life is somewhat subject to chance—for argument's sake, let's say that life is a "one-in-a-billion" event, or that life originates on one out of a billion wet planets—there would be trillions of examples of independently evolved life in the universe. We have only one example (Earth) of a wet planet on which life evolved out of only one wet planet visited (Earth). So, it's impossible to estimate the probability of a wet planet to spark life, but based on our data, life would have to be unimaginably rare for us to be alone in the universe. Further, if we were to find one more example of independently evolved life on one other planet (or moon), then I think it would become very difficult to argue that we are alone in the universe. Finding one more example would, in my view, eliminate the possibility that we are alone.

What is life?

Life is easier to characterize than it is to define. The key characteristics are: Living organisms are compartmentalized (i.e., they are sequestered from their environment in little cells by little walls made of carbon-containing molecules); they take energy from the universe to maintain the order that allows that compartmentalization; they reproduce; and they evolve, generation after generation, as they do so.

That textbook characterization fails to define life per se, but we use it as a tool to guide our thinking of how to go about answering the question "Are we alone in the universe?"

My purpose is not to answer that question (indeed, we can't answer that question definitively today); rather, I would like to build a case for spending the time, money, and effort to go out and find life elsewhere in the solar system in the next few decades. My argument has three parts: i) if we were to discover life in the solar system, we may likely conclude that the universe is flooded with life; ii) the odds seem good that we will find life somewhere in this solar system, based on how long life has existed on this planet and how prevalent it has become here; iii) aside from the metaphysical value of demonstrating that we are not alone in the universe, there are, I believe, inevitable economic benefits of going out and finding other examples of life, and those economic benefits could be massive.

The kind of life I discuss here is not "intelligent" life. (For reasons that I exclude, many believe that time is too short and the distances between stars too long for us to expect ET to call us up on the radio.) I discuss, instead, the "simplest" examples of organisms that eat, grow, and reproduce. On Earth, bacteria and archaea represent the prime examples: these microbes are single-celled organisms that use diverse sources of energy (usually sunlight and/or chemicals naturally present on our planet) to grow and reproduce. Each single-celled microbe is compartmentalized from its environment by oily walls and membranes that hold the DNA, and that DNA encodes all of the RNA and enzymes it needs to make copies of itself and carry out all of the networks of chemical processes involved in eating, growing, and reproducing. Those simple examples are complicated enough to be considered life, and finding analogous examples on another heavenly body (e.g., Mars or Europa) satisfies the requirement.

While the time and distances might not ever allow intelligent species to contact each other using light signals, the timelines for the development of life here on Earth are encouraging. It apparently did not take long for life to spark on Earth—not long in the grand scheme of things. The Earth is roughly 4.55 billion years old, but the oldest evidence we have for life is, depending on who you believe, between 3.7 and 4.0 billion years old. Importantly, during the first half billion years after Earth accreted, it was not at all the kind of place where one would expect life to spark and take hold. First, the crust was not completely cooled—molten rock comprised much of the surface of the planet. Second, asteroids and comets constantly bombarded the Earth in sterilizing collisions. In fact, as best as we can determine, Earth was not habitable until just over four billion years ago, but life was prevalent enough to be trapped in the fossil record by 3.7 to 4.0 billion years ago. Thus, from the time that Earth was habitable to the time that life took hold amounted to as little as a few million and not longer than few hundred million years. In either event, the span was not long on geologic time scales, and that short time scale (to me) reflects how easy it is for life to spark. Life happened quickly on Earth (within millions of years of it being habitable), and given an analogous environment on another planet or moon with the right conditions (presumably, liquid water and an energy source), I like the chances of it happening elsewhere, too.

When does it start?

Without a doubt, the answer to the question, "How does life start?" is "We don't know." What we do know is that simple life seems to have started within millions of years of there being both liquid water on the surface of the planet and abundant energy sources (sunlight, lightning, volcanism, meteor strikes, tides in the ocean and even in the soil because the moon was closer then, to name a few). There was also one additive crucial for life as we know it: carbon-containing molecules. We think the young Earth was bathed both in water and in carbon-containing molecules like amino acids, simple sugars, and nitrogenous bases—all similar to the components we find in every example of life on Earth today. This belief arises because, now 4.5 billion years later, we find water and carbon-containing molecules in the countless meteorites and comets in this solar system. The young Earth, four billion years ago, was constantly showered with organic material and water as abundant meteorites and comets rained in from space.

While no one today can credibly provide a detailed picture of how a simple organism, capable of self-replication, could spontaneously arise from this primordial soup, many believe that the crucial components of that soup are water, energy, and carbon-containing molecules. I go one step further to believe that if life happened quickly on Earth, it likely also happened more or less quickly on other heavenly bodies that had similar components. That is, admittedly, quite speculative, but it forms the basis of my optimism of finding evidence for life (either past or present) elsewhere in this solar system because we have, through the tireless efforts of our robotic ambassadors, derived evidence for there being water, energy, and carbon-containing molecules elsewhere in this solar system. Furthermore, the conditions here on Earth under which we currently find life make a strong case for finding evidence of life elsewhere in this solar system. The numbers and sizes of planets found orbiting other stars in our galaxy indicate that our solar system is not at all unique in the cosmos.

Life on Earth makes a strong case for life in this solar system.

Considering that humanity has scoured the Earth and found life everywhere, *everywhere* it has looked, I think it likely that life exists elsewhere in this solar system. Here on Earth, microbes thrive in places where plants and animals could not. These fascinating examples form the basis of part of my optimism: below are descriptions of two extreme examples of where life has evolved to thrive on Earth and how those two places are not all that dissimilar to other heavenly bodies of this solar system.

Microbes in Salt in the Atacama Desert: Life thrives in places where there is almost no liquid water (at least, not much liquid water). The Atacama Desert of Chile is one such place: annual rainfall there is measured in millimeters per year. One could be forgiven for expecting the Atacama Desert, which is inarguably the driest place on Earth, to be lifeless. Remarkably, we have discovered whole ecosystems of microbes living in salt-laden soils just a few meters below the surface of the Atacama, where these industrious little organisms have adapted an ingenious approach to thriving in such an arid environment. It turns out that parts of the Atacama subsurface comprise salt deposits (like, for example, halite or rock salt) that are hygroscopic and deliquescent. Hygroscopic salts absorb water vapor from the air, and deliquescence occurs when a salt absorbs so much water vapor that drops of liquid water actually condense on the surface of the solid salt. Even though it almost never rains on the Atacama Desert, these salt deposits in the soil reap the sparse water vapor in the air, which deliquesces on the surface of the salt, forming a vanishingly thin sheet of liquid water—the thickness of that little pool of water on the surface of salt crystals can be about the same as the thickness of a hair on your head. To a microbe, however, that is an ocean of water, and those little pools, meters below the surface, support copious tiny ecosystems of microbes that eat, grow, and reproduce. We currently have no evidence that any such thing could occur on Mars, but the example certainly makes me excited about the possibilities of how life, had it ever established a foothold on a young, wet Mars, might have evolved a way to survive the desiccation of the surface of that planet.

Buried, Frozen Lakes Teem with Life: Near the poles of our planet, we find a lot of water ice. That ice, in some places, is miles thick. Occasionally, in divots in the land beneath that ice, there are small lakes of liquid water covered by a couple of miles of ice. We have drilled through the ice to sample water from these frozen lakes (Lake Vostok and Lake Whillans in Antarctica), which have been completely cut off from the life elsewhere on the planet for millions of years. The amazingly surprising result, to me, is that the buried lakes *teem* with life. To exist, thus, life does not need ample sunlight, like that in which we bask on the surface of the Earth. Life does not need the oxygen in the air that we breathe. Life can exist when just liquid water and a chemical source of energy are present. These organisms appear to feed off of carbon dioxide, iron, ammonia, and sulfur in the detritus. They thrived in liquid water that hovers at the freezing temperature, and for millions of years they never, ever saw sunlight, nor did they have access to oxygen. These little microbes, nonetheless, continued to procreate and evolve on their own for eons, and reap the energy of the carbonaceous material at the bottom of these frozen lakes.

The ubiquity of life on Earth supports the expectation that we will find it elsewhere. In our solar system, there are places (places we could explore thoroughly within just a few decades) where conditions are not completely dissimilar to those described above. We already suspect that Mars and Europa (the subjects of the next sections) have or have had the three crucial components for life to spark: liquid water, an energy source, and carbon-containing molecules. So far, our robots have hardly looked; to find life elsewhere, we need to send well-equipped emissaries to look hard.

Mars, Life on the Red Planet: As we study Mars, we collect compelling data that indicate that it was wet sometime during its 4.5-billion-year existence and that liquid water may still be present there today. The Curiosity probe chemically analyzes rocks on the Red Planet and these analyses indicate at least two things: i) the presence of all the important elements of life (as we know it on Earth); and ii) that at some point in the past, Mars was wet. Our observation of dead volcanoes there further indicates that at some time in the past, Mars provided an energy source to spark life. It may be reasonable to think that conditions on Mars four billion years ago might not have been all that different from those on Earth at that time. Since then, Mars has lost much of its water, but there is evidence that liquid water still exists near the surface. The Mars Reconnaissance Orbiter provides images we interpret as mountain streams growing and shrinking as the seasons shift on the surface. Even if Mars were only wet and geologically active for a few hundred million years—roughly the amount of time that it took for life to spark and take hold on Earth—the possibility exists that life sparked there, too. If life can survive in thin slivers of water on salt buried under the Atacama Desert, why would one think that it could not do the same on Mars?

Europa, Life Under the Ice: The probes we have sent zooming past Europa lead us to believe that this moon of Jupiter is a crust of ice on top of an ocean of liquid water, although we know nothing about the geologic activity (the energy source) of the presumed rocky surface at the bottom of that ocean. There is, however, a candidate for an energy source that might have provided the spark on Europa: tidal friction. As Europa orbits Jupiter, the gravity of the gas giant pulls and squeezes the ice and water of the tiny moon. That cyclic pulling and squeezing constantly stretches and crushes ice to the point where we see changes in the cracks and fissures of the surface ice over time. Even recently, the Hubble Space Telescope captured pictures of what is believed to be water spraying out of these cracks and fissures into space. We know that frozen lakes on Earth can harbor abundant life—we do not know if tidal friction alone provides enough of an energy source to spark life. Nonetheless, the combination of liquid water and a constant energy source, to me, make a compelling case for one to expect to find microbes (at least) on Europa. The technical challenges of getting a robot there would be enormous, of course, but perhaps not as difficult as getting there and burrowing through unknown depths of ice. The tectonic activity of the ice crust seems to indicate that water is cycled from the ocean to the surface. Finding microbes on Europa could be as simple as sending the right instruments to land on the ice and start to search.

Why go?

To justify the effort of going to find life elsewhere in the solar system, many cite the innate curiosity of humanity. I empathize; I personally need no other justification than simply establishing that we are not alone in the universe. Unfortunately, i) I do not have the resources to make it happen; and ii) the governing bodies of our society are faced with tackling myriad problems that are more important to the perpetuation of humanity—in the short term, at least—than satisfying our curiosity. To me, there is some likelihood that a corporation will ultimately need to bear the cost (and the risk) to reap the reward of finding life elsewhere in this solar system.

Return on Investment: To send people to Mars might cost on the order of hundreds of billions of dollars; to send a fleet of robotic probes with all the needed instruments would cost perhaps a tenth of that number. It is inconceivable to me that we will send people to Europa in my lifetime. In that case, robots are the only candidates to go there, and that could cost on the order of ten billion dollars per mission. Simply finding life, however, probably would not justify those costs to any corporation that exists today. To justify those costs they would need the potential for enormous returns on investment. We often cite examples of technology spawned by the Apollo and Shuttle programs (solar power, medical diagnostics, the artificial heart, etc.), but those programs were built at a time when we were just taking our infant and juvenile steps into space. Although assembling the technology to get there might yield some profitable breakthroughs for the funding entity, much of that technology already exists. We need a bigger reason. We need the possibility for grander returns.

I think those returns could come in the form of understanding how life works elsewhere. Life evolves to solve problems, and it does so with chemistry. Life on this planet evolved to solve the problem of living without sunlight. It evolved to solve the problem of living with minuscule amounts of water. It evolved to live without oxygen. And, life reaps water and energy from this Earth in countless more ways than the examples cited above. We have only, in the last few decades, begun to study the chemistry that life on Earth employs to solve these problems. Life elsewhere will have distinct solutions, and by studying life elsewhere, I believe, we will learn new ways to solve problems of reaping energy and water from our planet. When I consider the problems that threaten the continued existence of humanity on this planet, I cannot help but conclude that discovery of life elsewhere would have massive effects on the global economy. The owners (in this scenario, the corporation who foots the bill) of such solutions—along with answering The Question of whether or not there is life elsewhere in the universe—would enjoy the opportunity to profit from employing those solutions here on this planet. The trade-off for those of us who want to know comes in the form of ensuring that those who foot the bill have the right to profit from their discoveries.

Will we know in my lifetime?

The magnitude of the effort needed to find life elsewhere, if it exists on Mars or Europa, probably represents an incredibly tiny fraction (on the order of 0.01%) of the collective resources of humanity. Only a supportive, global consensus, however, could unite to tap the resources needed. Building that consensus requires not only a shared understanding of the costs, risks, and rewards, but also an agreement on the importance of such an endeavor to society on a grand scale. I hope (but I am not overly optimistic) that we will find life elsewhere during my lifetime. I see some hope in the confluence of three societal trends. First, privately funded space flight went from the drawing board to a reality (SpaceX, Orbital Sciences Corporation, etc.) in one decade. Second, our functioning satellites have shrunk in size, weight, and cost (cubesats). Third, our interconnected society is rapidly changing the ease with which good ideas build financial support through crowdfunding (for example, Kickstarter). One romantic vision could emerge as a fleet of low-cost probes, launched by a private corporation, funded by millions of interested individuals, racing to answer one of the biggest questions humanity has ever asked. As unlikely as that may sound, I think it a more likely scenario than NASA finding life elsewhere in the next several decades. Nothing about the current pace of our governmentally funded space program fills me with hope.

A positive identification of life elsewhere would be the greatest singular accomplishment of the system of life that sparked here on Earth some four billion years ago. The distinct possibility yet remains that we are alone, and that we might not ever find life elsewhere in this solar system, either in my lifetime or in another. Even in that disheartening case, establishing that Earth hosts the only life in this solar system would not be a failure if it were to yield some global appreciation of just how unique and precious and impossible life is. I waste little time worrying about that scenario, however. Life happened here on Earth almost as quickly as this planet became habitable, and today it exists almost everywhere we have looked. And, perhaps, in a place we have not.

When We're in Space

Sarah Blake

He is the planet on its side.
I'm waving at him

when he turns from me,
and I'm kissing him

the rest of the time.
I know, I know I'm not

the anti-cyclonic storm,
driven by his planet-heat.

Pictures of her are moving—
pictures of her like an eye.

But God, if one shooting star
doesn't follow another.

Peak Radiant

Linda Gallant

The Outer Banks, North Carolina, 12:30 (ish) a.m., August 12, 2013

We're spread out in chairs and on beach towels watching our friend Phil scamper around the beach wearing a fisherman's headlamp. It's supposed to help him see his catch in the dark, which it does, but it also affords him the opportunity to shine an incredibly bright light in our faces while we drink and wait for another bright light to appear on the dark beach. True dark, the kind of dark that exists far from city glare and phalanxes of brightly lit shore homes.

I'm on vacation with my boyfriend's friends, many of whom are scientists like him. We're on what feels like a beach on another planet, one free of the pollutants that prevent us from seeing the sky in crystalline full. We are lucky; there are only a few clouds, which we can also clearly see in a sky so black and limpid. The guys know more about the details of the Perseids meteor shower we're about to witness than I do, and I start half-listening to them talking about the Swift-Tuttle comet when the first burst—*burst*—of light illuminates the horizon, so much closer to the ocean than I'd imagined the meteor would be. Its movement is lateral—the intense yellow blur burns across the horizon line for a few distinct seconds of pure surreality. The shocked silence is cut by primal, joyful hollering up and down the beach. We're not the only ones out to see the show.

...

According to the New Jersey Department of Environmental Protection, the nearest air quality station to my hometown on the Jersey Shore is located on Monmouth University's campus in West Long Branch. There is a link on the website that connects you to Camnet,[1] a real-time air pollution and visibility monitor displaying views of both rural and urban areas caught on "hazecams" (designed, as one would assume, to show evidence of obscuring pollution). You can compare "ideal views" to current conditions in Newark or Boston or Brigantine. The website helps you interpret the visibility data collected each day by the cams:

> *Is it a hazy day?* Haze is relatively uniform at the horizon but tends to diminish slightly at higher elevations. Look at the levels of man-made pollutants including fine particles, black carbon, and ozone. Also, note the relative humidity. Haze often occurs on hot, humid summer days with medium or high levels of fine particles, ozone, and sometimes black carbon.
>
> *Is it a brown cloud day?* A brown cloud appears to envelop the scene but quickly thins out at higher elevations. Brown clouds tend to occur on calm winter mornings during rush hour traffic. Look at the particle and black carbon levels—they are usually high. Ozone will be low and relative humidity may vary.[2]

After living in New York and Philadelphia over the course of the last six years, I love going home to the shore, but not necessarily for the summertime traffic-lotion-traffic-beach-lotion-traffic ritual I resent more and more each year. The trips home I relish most, pale-faced contrarian that I am, happen during the fall. I slam my car door shut after the drive and sharply breathe in the air that smells like burning leaves. I see degrees more of stars in the night sky than I do in Philadelphia, having left the light/actual pollution behind. I look straight up and recognize constellations, happy to be visiting my folks and breathing carefree lungfuls of comparatively clean air.

Taking a trip to the beach at night during the off-season is even more of a studies-in-stark-contrast delight. I like looking at a deserted stretch of sand after witnessing thousands of tourists (i.e., "bennies"[3]), slick with sweat and layers of sunscreen, jostling for position every summer when the town's population triples. The pale sand and blue-black sky meet the terrifyingly vacant Atlantic, by turns dark and frothy white depending on the tide and wind speed. The stars are even more plentiful here, away from the competitive strip of twinkling commercial properties. The stars and half moon illuminate the shuttered shore houses, terrifyingly dark and vacant in their own way, left for a kind of dead by the wealthy families who flip the switch and convert a house into a home for three or four months during the year.

...

On the outskirts of another shore community, the drunken scientists know that the blips and bursts of light are just pieces of the Swift-Tuttle comet's tail burning up in Earth's atmosphere as we pass through the swath. But they are as stupid-giddy as those of us who can only describe what we see as impossibly far away fireworks, sent up by the ancient inventors themselves, from a hilltop halfway around the world in celebration. We keep drinking and dancing and triumphantly flashing Phil's headlamp until the shower begins to wane. We walk away from the spot we'd staked out, careful to pick up the solo cups we tossed on the first sighting, unable to hold on to anything but each other.

The Anatomy of Man's Body

AS GOVERN'D BY THE TWELVE CONSTELLATIONS

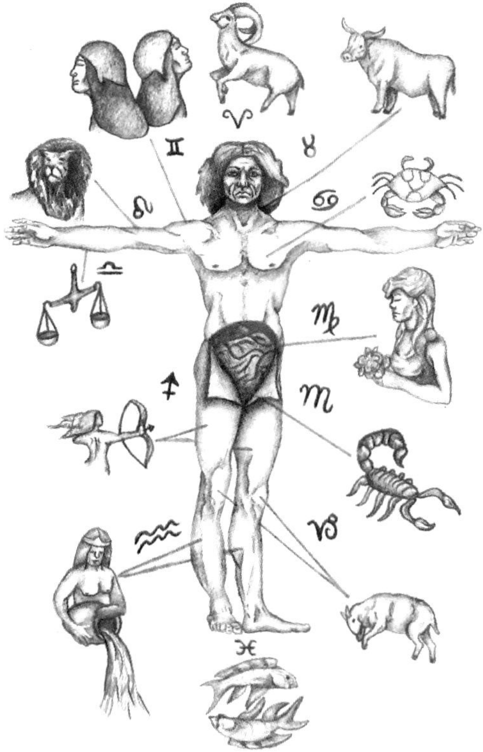

Alexandra Montgomery

A 2014 Guide to Meteor Showers

"All the firmament seemed enveloped in splendid fireworks, as if every star in the broad expanse had been hurled from its course..."
—Parley P. Pratt[4]

Lyrids

Radiant: Constellation Lyra *(Latin for "lyre")*
What we see: Debris from comet C/1861 G1 Thatcher
Dates: April 6-25
Peak night: April 21-22
Best viewed from: Northern Hemisphere, at dawn

Eta Aquariids

Radiant: Constellation Aquarius *(Latin for "water-carrier")*
What we see: Debris from Halley's Comet
Dates: April 19-May 26
Peak night: May 5-6
Best viewed from: The southern tropics

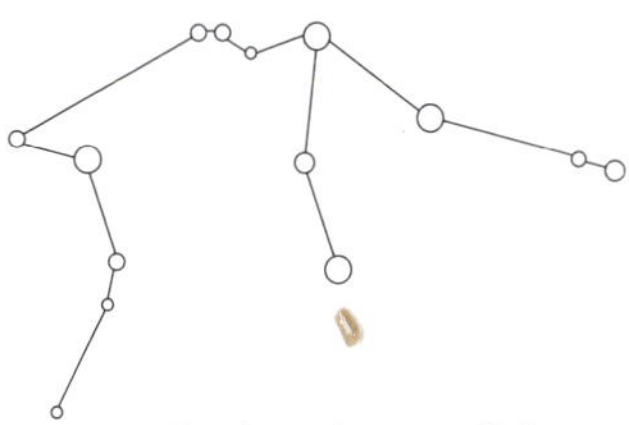

Delta Aquariids

Radiant: Constellation Aquarius
What we see: Debris from Marsden and Kracht Sungrazing comets
Dates: July 21-August 23
Peak night: July 27-28
Best viewed from: The southern tropics

Alpha Capricornids

Radiant: Constellation Capricornus *(Latin for "horned goat")*
What we see: Debris from comet 169P/NEAT
Dates: July 11-mid-August
Peak night: July 28-29
Best viewed from: Either side of the equator (but preferably the southern tropics)

Perseids

Radiant: Constellation Perseus *(named for the Greek hero Perseus, slayer of Gorgon Medusa)*
What we see: Debris from Swift-Tuttle comet
Dates: July 13-August 26
Peak night: August 11-12
Best viewed from: The Northern Hemisphere (This shower is widely considered to be the most popular and prolific shower among those listed here)

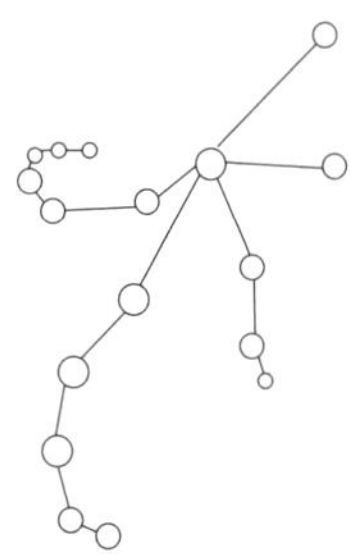

Orionids

Radiant: Constellation Orion *(named for the Greek hero and huntsman Orion, home of the red supergiant Betelgeuse)*
What we see: Debris from Halley's Comet
Dates: October 4-November 14
Peak night: October 21-22
Best viewed from: Either side of the equator

Southern Taurids

Radiant: Constellation Taurus *(Latin for "bull")*
What we see: Debris from the comet Encke
Dates: September 7-November 19
Peak night: October 8-9
Best viewed from: The southern tropics

Northern Taurids

Radiant: Constellation Taurus
What we see: Debris from the comet Encke
Dates: October 19-December 10
Peak night: November 12-13
Best viewed from: The Northern Hemisphere

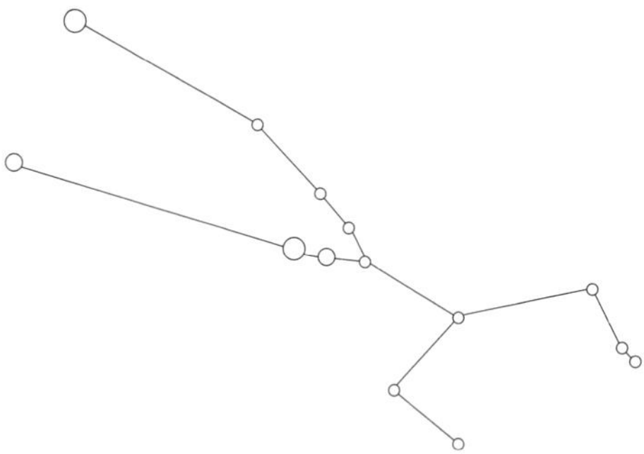

Leonids

Radiant: Constellation Leo *(Latin for "lion," specifically the lion killed by Hercules in Nemea as one of his labors)*
What we see: Debris from comet 55P/Tempel-Tuttle
Dates: November 5-30
Peak night: November 17-18
Best viewed from: The most prolific showers have been documented as occurring over the Americas and Europe and were so spectacular that they prompted greater scientific inquiry into meteor showers in the nineteenth century. Sadly, the next significant display is not predicted to appear until 2099.

Geminids

Radiant: Constellation Gemini *(Latin for "twins," named for brothers Castor and Pollux)*

What we see: A prolific and intense shower originating not from a comet but from object 3200 Phaethon, an asteroid

Dates: December 4-16

Peak night: December 13-14

Best viewed from: The Northern Hemisphere from 10 p.m. onward (a rare shower visible before midnight)

Ursids

Radiant: Constellation Ursa Minor *(Latin for "smaller bear," the tail of which is the Little Dipper)*

What we see: Debris from comet 8P/Tuttle

Dates: December 17-23

Peak night: December 21-22

Best viewed from: The Northern Hemisphere during the late morning hours

Dates courtesy of the American Meteor Society

Weather Glossary

In Preparation for what Will certainly be increasingly Erratic Weather Patterns in our future, the Editor of this Almanac encourages you, Courteous Reader, to become familiar with this Miscellany of Terms, Concepts, and Descriptors.

1. **al·be·do** *n.* The percentage of light reflected by an object. Snow covered areas have a high albedo (0.9 or 90%) due to their white color.[1]

2. **an·vil zits** *n.* Frequent (often continuous or nearly continuous), localized lightning discharges occurring from within a thunderstorm anvil.[2]

3. **bar·ber pole** *n.* A slang term used by storm spotters to describe a thunderstorm updraft with a striking visual appearance: cloud striations that are curved similar to the stripes of a barber pole.[3]

4. **bear's cage** *n.* A region of storm-scale rotation, in a thunderstorm, which is wrapped in heavy precipitation. This area often coincides with a radar hook echo and/or mesocyclone, especially one associated with an HP storm. The term reflects the danger involved in observing such an area visually, which must be done at close range in low visibility.[4]

5. **bron·tide** *n.* A low muffled sound like distant thunder heard in certain seismic regions especially along seacoasts and over lakes and thought to be caused by feeble earth tremors.[5]

6. **Chi·nook wind** *n.* A warm, dry wind blowing down the eastern slopes of the Rockies. This wind is responsible for the most

dramatic temperature rise on record —a 49-degree temperature rise that occurred—in just 2 minutes—in Spearfish, SD, in September 1943. Chinook translates as "snow eater," and the wind often forms after a very intense cold spell.[6]

7. **con·gest·us** (or **cumulus congestus**) *n.* A large cumulus cloud with great vertical development, usually with a cauliflower-like appearance, but lacking the characteristic anvil-shaped top of a cumulonimbus (Cb).[7]

8. **cry·o·sphere** *n.* One of the interrelated components of the Earth's system, the cryosphere is frozen water in the form of snow, permanently frozen ground (permafrost), floating ice, and glaciers. Fluctuations in the volume of the cryosphere cause changes in ocean sea level, which directly impact the atmosphere and biosphere.[8]

9. **dea·sil** *adv.* Right-handwise, sunwise, clockwise— used especially of the Masonic rite of circumambulation and also of Masonic floor work that is clockwise. The opposite is **widdershins**, meaning anticlockwise.[9]

10. **glo·ry** *n.* An optical phenomenon that resembles a rainbow halo around an object's shadow when it's directly between the sun and a cloud.[10]

11. **grau·pel** *n.* Ice pellets, generally two to five millimeters in diameter. Graupel are oftentimes mistaken for hail, but a simple way to tell the difference is to touch a graupel ball. Graupel will typically fall apart when touched or when they hit the ground, whereas hail is very hard as a result of being formed when layers of ice accumulate, and is associated with thunderstorms.[11]

12. **ha·boob** *n.* A dust or sandstorm caused by the downdraft of a desert thunderstorm.[12]

13. **ha·lo·car·bons** *n.* Compounds containing either chlorine, bromine or fluorine and carbon. Such compounds can act as powerful greenhouse gases in the atmosphere. The chlorine and bromine containing halocarbons are also involved in the depletion of the ozone layer.[13]

14. **knuck·les** *n.* Slang for lumpy protrusions on the edges, and sometimes the underside, of a thunderstorm anvil. They usually appear on the upwind side of a back-sheared anvil, and indicate rapid expansion of the anvil due to the presence of a very strong updraft. They are not mammatus clouds.[14]

15. **niv·e·ous** *adj.* Of or relating to snow: resembling snow (as in whiteness).[15]

16. **out·gas** *v.* **-gassed, gassing**. Release of gases into the atmosphere from hot, molten rock during volcanic activity; thought to be the origin of most atmospheric gases.[16]

17. **Pan·han·dle hook·er** *n.* No, not that kind. An infrequent storm system that originates in the Panhandle region of Texas and Oklahoma, initially moves east and then "hooks" or re-curves more northeast toward the upper Midwest or Great Lakes region.[17]

18. **phe·nol·o·gy** *n.* The timing of natural events, such as flower blooms and animal migration, which is influenced by changes in climate.[18]

19. **po·lar vor·tex** *n.* A circumpolar wind circulation which isolates the Antarctic continent during the cold Southern Hemisphere winter, heightening ozone depletion.[19]

20. **res·i·dence time** *n.* The average time spent in a reservoir by an individual atom or molecule. With respect to greenhouse gases, residence time refers to how long on average a particular molecule remains in the atmosphere.[20]

21. **sink** *n., v.* Any process, activity or mechanism which removes a greenhouse gas, an aerosol or a precursor of a greenhouse gas or aerosol from the atmosphere.[21]

22. **snow·poc·a·lypse** *n.* A term coined in the latter part of the first decade of the 21st century. It is used to describe a weather phenomenon where a region or regions is/are struck by massive and crippling snow storms back to back. Unfortunately (or fortunately, depending how you look at it) the term is often mispronounced by hair-helmeted, stuttering, stumble-bummed weather forecasters and anchor persons. If the use of this word follows past trends, soon every fucked up weather pattern will be renamed with the suffix "-pacalypse." Just as every dicked up scandal or back door deal was standardized by the suffix "-gate" after the "Watergate" scandal.[22]

23. **sno·ver·kill** *n.* A weather condition consisting of heavy snow after a snowpocalypse. This storm follows only days after the snowpocalypse which the city has yet to overcome from, perpetuating the mass panic and sense of despair.[23]

24. **su·per snow·va** *n.* A blizzard of such magnitude and intensity that it could explode, create a black hole, and destroy the solar system.[24]

Contributors

Sarah Blake lives just outside of Philadelphia with her husband and son. In 2013, she received a Literature Fellowship from the NEA for her poetry about Kanye West. Her debut collection of poetry, *Mr. West*, is forthcoming from Wesleyan University Press later this year.

Kristy Caldwell makes visual stories. The illustrations in the *Mars* graphic novel preview were originally created for the 2013 Superhero Clubhouse/Matchboxarts production of *Mars! A Play About Mining*, which included story and text adapted from Tom Coiner's graphic novel. Most recently she wrote and illustrated *Bachi and the Rooftop of the World*, which will be released in 2014 on the Imaginari app. She received the Award of Excellence from *Communication Arts*, and her work has also been recognized by *American Illustration* and *Creative Quarterly*. She is co-owner of the quarterly publication *Carrier Pigeon: Illustrated Fiction and Fine Art*. Kristy was brought up in Louisiana and now lives and works in New York. Find more of her work at www.kristycaldwell.com.

Tom Coiner is an actor and writer who lives in New York City. He's a graduate of Kenyon College and the National Theatre Conservatory. He was born in Marlborough, MA, the town which he credits for his taciturn demeanor. *Mars* was commissioned by Jeremy Pickard and the Superhero Clubhouse, a New York-based theatre company, for use as the basis for their play *Mars! A Play about Mining.*

Nic Esposito is a writer, urban farmer, and founder of The Head & The Hand Press. His contribution to the *Almanac* is from a forthcoming novel he is working on inspired by his obsession with Justus von Liebig—a man who substantially changed the course of human history without many humans even realizing it. He lives on his urban homestead in the Kensington neighborhood of Philadelphia with his wife, Elisa.

Jonathan Farbowitz is a computer programmer, freelance copyeditor, and sometimes poet living in West Philadelphia. He studied film and philosophy at Vassar College, and has worked as a film editor, a reading tutor, and a bicycle safety educator.

Linda Gallant is an editor who lives in Philadelphia with a poorly behaved dog. She holds a master's degree in English and American Literature from New York University and helps run The Head & The Hand Press's Writers' Workshop.

Sara Graybeal is a Philadelphia-based youth educator, writer, and spoken word artist. Her full bio and links to other works can be found here: http://backyardwriters.org/backyarders/sara-graybeal/.

Sarah Grey is a freelance writer and editor at Grey Editing LLC (www.greyediting.com; @greyediting) in the Fishtown section of Philadelphia. Her editing practice specializes in history, sociology, media studies, and geopolitics, particularly the Israel-Palestine conflict. She also enjoys working on fiction and poetry, as well as translated work. Her writing has been published in the *Rust Belt Rising Almanac, Grid, Motivos, Monthly Review,* and *International Socialist Review,* among others. When she doesn't have her head buried in a manuscript, she enjoys playing Chutes and Ladders, attending protest marches, and hosting spaghetti-and-meatball dinners with her four-year-old daughter Lucia and husband Joe.

Wintfred Huskey lives and works in Philadelphia. His novel *Blowin' It* is forthcoming from The Head & The Hand Press this summer.

Ray Jones is a graduate of SVA'S MFA Illustration program (2009), Ray Jones works as a freelance illustrator, designer and teacher in NYC. Clients include: Sony UK, *Esquire Russia*, the *New York Times*, the *Seattle Stranger*, QN5 Music and more. Contact Ray for work at: ray@jonesray.com.

Angela Miles is a graphic designer, typeface enthusiast and stained glass artist living and working in Philadelphia, Pennsylvania. She holds a Bachelor of Sciences degree in Graphic Design Communications from Philadelphia University and specializes in editorial design & illustration, typography, concept development, content mapping and brand identity. Angela currently is employed by the Community College of Philadelphia and runs her freelance business, Carpe Diem Construction Co., working with and for musicians, theater groups, independent publishing companies, non-profits, small businesses and other artists & designers. www.carpediemconstruction.com.

Alexandra Montgomery is currently finishing her BFA degree in Sculpture, with a teaching certification for grades K-12, at Tyler School of Art in Philadelphia. She primarily works as a mold maker and fabricator, working as both an object maker and an installation artist. She worked as an intern with mural artists Donald Gensler and Phillip Adams in 2010, and her portrait is featured in their City Hall Concourse Mural in Philadelphia. More of Alexandra's work can be viewed at: cargocollective.com/alexandra_montgomery.

Christine Neulieb is a writer, freelance editor, and fledgling publisher. She's been writing for as long as she can remember—you can find links to her work at www.christineneulieb.com or follow her @neulita. She started freelance editing in December of 2012, and now operates Lantern Editorial (www.lanterneditorial.com) out of Independents Hall in Old City, Philadelphia. While there, she met graphic designer Amanda Thomas, and together they have co-founded a small indie publishing house, Lanternfish Press (www.lanternfishpress.com). Lanternfish's first title, *The Legend of Sherlock Holmes*, was successfully funded via Kickstarter and is due to hit shelves in late March. Christine also holds a doctorate in ancient Greek philosophy of literature and likes to get super nerdy about theory. She's lived a lot of places, but Philadelphia is her adopted hometown.

Miriam Seidel is a writer and curator in Philadelphia. She's written librettos for two operas: *Violet Fire*, about Nikola Tesla, and *Judgment of Midas*, based on a story from Ovid's *Metamorphoses*. Her short fiction has appeared in *Exquisite Corpse*, *Washington Square* and elsewhere, and her articles and reviews have appeared in *Art in America*, the *Philadelphia Inquirer*, *Dance*, and many other publications. Her two (unpublished) novels touch on the science fiction fan world and the Alexandria Library. More information is at www.miriamseidel.com.

Before she became a member of the South Jersey Poets' Collective, ***Jacalyn Shelley*** was an insurance executive and business systems consultant. She earned a master's degree in History at Monmouth University and taught there for two years. She now leads an extension literature class affiliated with The Richard Stockton College.

Eric Smith is the co-founder of Geekadelphia and the Philly Geek Awards, and the social media manager at Quirk Books. He's the author of *The Geek's Guide to Dating*, and his debut YA novel *Inked* is due out in the fall of 2014 with Bloomsbury Spark. He's written for a number of places, including The Huffington Post, Boing Boing, BuzzFeed, BookRiot, and The Bygone Bureau.

Phillip Snyder earned his bachelor's degree in chemistry at Allegheny College in Meadville, PA, and his Ph.D. in Organic Chemistry at Duke University in Durham, NC. During his postdoctoral work at Harvard University in Cambridge, MA, Phil studied biophysics and, for a brief period, the chemical origins of life on Earth. He is currently an employee of Vertex, Pharmaceuticals Inc. in Boston, MA, where his group applies physical chemical approaches to the discovery and development of new medicines. In his spare time, Phil enjoys reading about cosmology and particle physics, fishing, and amateur astronomy.

Christopher Wink is the co-founder of Technical.ly, a network of local technology news sites, and lead organizer of Philly Tech Week, Baltimore Innovation Week, and other events that bring smart people together. Previously, he worked for a homeless advocacy nonprofit and was a freelance reporter. Wink maintains a creative writing portfolio, having most recently been anthologized in the *Rust Belt Rising Almanac*, and tweets @christopherwink. The Temple University alumnus is a bicycle commuter and resident of the Fishtown neighborhood of Philadelphia.

Alex Vuocolo is a reporter for Generocity.org and freelance writer on urban affairs in Philadelphia. He is also a proud college dropout and Philadelphia resident, two qualifications that imply cheap rent and minimal debt.

Kathryn Zazenski's work alludes to phenomenological experience by way of every day objects and interactions. Using video, projection, installation, sound and still images, she opens a dialogue centered on communication and expectation that splinters into ideas about truth, faith, seduction, and the balance of dualities like attraction and repulsion. Zazenski holds a BFA from Kutztown University of Pennsylvania and an MFA in Sculpture from Cranbrook Academy of Art. www.kathrynzazenski.com.

Acknowledgments

The creators of *The Asteroid Belt Almanac* would humbly like to thank...

Claire Margheim, true-aiming layout scout.

Angela Miles of Carpe Diem Construction Co., design sherpa extraordinaire.

Our illustrious staff: ***Nic Esposito***, ***Linda Gallant***, ***Claire Margheim***, and ***Kerry Boland***.

Erin O'Neil, intrepid publicity and marketing intern.

Joe Lerro and ***Lisa McHenry Bendel***, steely-eyed copy editors.

Mike Baccam, editorial assistant and literary homing beacon.

Carl Cheeseman, composer and friend.

Karen Kircher and Versa Press, the printer's printer.

Technical.ly and ***Geekadelphia*** for their bleeding edge ingenuity.

Elizabeth Fuller of the Rosenbach Library, guide to the almanacs of yore.

William Penn, for his grand experiment.

Benjamin Franklin, for giving us the model almanac.

Our friends and family who invested their time, love, and money into this project.

Our wonderful writers and artists whose work fills these pages.

Stargazers of the past, present, and future.

Notes to Pages 39-47

1. Susan Squier, "Negotiating Boundaries: From Assisted Reproduction to Assisted Replication," in *Playing Dolly: Technocultural Formations, Fantasies, and Fictions of Assisted Reproduction*, edited by E. Ann Kaplan and Susan Squier (New Brunswick, NJ: Rutgers University Press, 1999), 104.

2. The term "uncanny valley" describes "why zombies are scary": robotics research has found that people react most negatively to things that look almost, but not quite, human. The website TV Tropes quotes C. S. Lewis's *The Lion, the Witch, and the Wardrobe*: "If you see something that...looks human and isn't, you keep your eyes on it and you feel for your hatchet" (TV Tropes, "Uncanny Valley," n.d., http://tvtropes.org/pmwiki/pmwiki.php/Main/UncannyValley).

3. Gillian Mohney, "3D Printer Makes Model Memento of Fetus for Parents," ABCNews.com, July 20, 2013, http://abcnews.go.com/blogs/health/2013/07/20/3-d-printer-makes-model-memento-of-fetus-for-parents.

4. Created to balance the right to free speech with patients' right to access medical care, "buffer zone" laws establish how close to a clinic's entrance (or to the homes of its employees) protesters are allowed to stand. Some also regulate threats, property damage, weapons possession, noise levels, and other forms of harassment against clinics and their patients and employees. The federal law is the Freedom of Access to Clinic Entrances Act; many state and local governments also have their own ordinances.

5. Guttmacher Institute, "State Policies in Brief: Requirements for Ultrasound," white paper, December 1, 2013, www.guttmacher.org/statecenter/spibs/spib_RFU.pdf.

6. Elizabeth Rosenthal, "American Way of Birth, Costliest in the World," *New York Times*, June 30, 2013, www.nytimes.com/2013/07/01/health/american-way-of-birth-costliest-in-the-world.html?_r=0.

7. U.S. House of Representatives, Committee on Government Reform, Special Investigations Division, "False and Misleading Health Information Provided by Federally Funded Pregnancy Resource Centers," July 2006, page i, www.chsourcebook.com/articles/waxman2.pdf.

8. Rachael Ninsiima, "Uganda: Philips Provides Ultrasound Technology for Rural Women," AllAfrica.com, July 16, 2013, http://allafrica.com/stories/201307171000.html.

Notes to Pages 158–161

1. A project of NECAUM, or the Northeast States for Coordinated Air Use Management.

2. "Camnet," accessed October 5, 2013, http://hazecam.net/camsite.aspx?site=brigantine#datadiv.

3. The etymology of this term is always in dispute. Some say it stands for "Bayonne, Elizabeth, Newark, New York." To residents of shore towns, it stands for "puke-stained sidewalk, nowhere to park."

4. "Spectacular Meteor Shower Might Repeat," accessed February 4, 2014, www.johnpratt.com/items/docs/lds/meridian/1999/meteors.html.

Notes to Pages 167-170

1. National Oceanic and Atmospheric Administration's National Weather Service Glossary, s.v. "albedo," last modified June 25, 2009, www.erh.noaa.gov/box/glossary.htm.

2. NOAA's National Weather Service Glossary, s.v. "anvil zit," last modified June 25, 2009, www.erh.noaa.gov/box/glossary.htm.

3. "Weird Weather Words," TravelChannel.com, accessed February 15, 2014, www.travelchannel.com/interests/weather/articles/unusual-weather-terms.

4. NOAA's National Weather Service Glossary, s.v. "bear's cage," last modified June 25, 2009, www.erh.noaa.gov/box/glossary.htm.

5. *Merriam-Webster OnLine*, s.v. "brontide," accessed February 25, 2014, http://unabridged.merriam-webster.com/unabridged/brontide.

6. "Weird Weather Words," TravelChannel.com, accessed February 15, 2014, www.travelchannel.com/interests/weather/articles/unusual-weather-terms.

7. NOAA's National Weather Service Glossary, s.v. "congestus," last modified June 25, 2009, www.erh.noaa.gov/box/glossary.htm.

8. "Glossary of Climate Change Terms," United States Environmental Protection Agency website, s.v. "cyrosphere," last modified September 9, 2013, www.epa.gov/climatechange/glossary.html#C.

9. *Merriam-Webster OnLine*, s.v. "deasil," accessed February 25, 2014, http://unabridged.merriam-webster.com/unabridged/deasil.

10. "Weird Weather Words," TravelChannel.com, accessed February 15, 2014, www.travelchannel.com/interests/weather/articles/unusual-weather-terms.

11. Ibid.

12. Ibid.

13. "Glossary of Climate Change Terms," United States Environmental Protection Agency website, s.v. "halocarbons," last modified September 9, 2013, www.epa.gov/climatechange/glossary.html#H.

14. NOAA's National Weather Service Glossary, s.v. "knuckles," last

modified June 25, 2009, www.erh.noaa.gov/box/glossary.htm.

15. *Merriam-Webster OnLine*, s.v. "niveous," accessed February 25, 2014, http://unabridged.merriam-webster.com/unabridged/niveous.

16. "Weird Weather Words," TravelChannel.com, accessed February 15, 2014, www.travelchannel.com/interests/weather/articles/unusual-weather-terms.

17. Ibid.

18. "Glossary of Climate Change Terms," United States Environmental Protection Agency website, s.v. "phenology," last modified September 9, 2013, www.epa.gov/climatechange/glossary.html#P.

19. NOAA's National Weather Service Glossary, s.v. "polar vortex," last modified June 25, 2009, www.erh.noaa.gov/box/glossary.htm.

20. "Glossary of Climate Change Terms," United States Environmental Protection Agency website, s.v. "residence time," last modified September 9, 2013, www.epa.gov/climatechange/glossary.html#R.

21. "Glossary of Climate Change Terms," United States Environmental Protection Agency website, s.v. "sink," last modified September 9, 2013, www.epa.gov/climatechange/glossary.html#S.

22. *Urban Dictionary*, s.v. "Snowpacalypse" entry by Armega Don, February 11, 2010, www.urbandictionary.com/define.php?term=Snowpacalypse.

23. *Urban Dictionary*, s.v. "Snoverkill" entry by LordMcWaffle, February 9, 2010, www.urbandictionary.com/define.php?term=snoverkill.

24. *Urban Dictionary*, s.v. "Super Snowva" entry by Elefante Borracho, February 2, 2011, www.urbandictionary.com/define.php?term=super+snowva.

Thank you to our "Early Adopters."

Your support helped create this very curious anthology.

Logan Axelson
Neil Barnes
Adrianna Borgia
Christine Chronis
Debra Finnegan-Suler
Michael Frederick
Douglas Gordon
Laura Heise
Andy Kahan
Patrice Leonetti
Allegra Love
Bridget Love
Kristan Menzel
Nicholas Mirra
Brian Nunziato
Crystal Ockenfuss
Sarah Peterson
Sarah Schoenholtz
Marc Schuster
Ricky Strickler
Asia Suler
Ashlie Thornbury
James Weissinger